Viruses, Pandemics, and Immunity

Arup K. Chakraborty and Andrey S. Shaw

illustrated by Philip J. S. Stork

The MIT Press
Cambridge, Massachusetts
London, England

This book was set in Stone Serif and Stone Sans by Jen Jackowitz. Printed and bound in the United States of America.

Library of Congress Cataloging-in-Publication Data

Names: Chakraborty, Arup, author. | Shaw, Andrey S., author.
Title: Viruses, pandemics, and immunity / Arup K. Chakraborty and Andrey S. Shaw.
Other titles: MIT Press first reads.
Description: Cambridge, Massachusetts : The MIT Press, [2021] | Includes index.
Identifiers: LCCN 2020033971 | ISBN 9780262542388 (paperback)
Subjects: MESH: Pandemics | Viruses--pathogenicity | Immunity | Vaccines | Virus Diseases--drug therapy
Classification: LCC RA644.V55 | NLM WA 105 | DDC 614.5/8--dc23
LC record available at https://lccn.loc.gov/2020033971

10 9 8 7 6 5 4 3 2 1

Viruses, Pandemics, and Immunity

To the victims and heroes of the COVID-19 pandemic

Contents

Prologue

Throughout history humans have contended with pandemics, and the written record since antiquity is replete with references to plagues, pestilence, and contagion. The end of the Middle Ages through the beginning of the Renaissance, a period of heightened creativity, was accompanied by a new freedom of movement around the Mediterranean and between Europe and Asia. But this increase in mobility was also linked to the spread of the plague. The great bubonic plague of the fourteenth century occurred from 1347 to 1352. It spread throughout Europe and the Middle East, killing about 100 million people. The plague likely started in China and then passed into Eastern Europe through the Silk Road. Once in Italy, it was able to spread widely across Europe and eventually crossed the Mediterranean to the Middle East.

This was not the first or last time the world suffered because of the plague. In AD 541, a pandemic began in Constantinople, the heart of the Byzantine Roman Empire, resulting in deaths equivalent to half the European population. The plague pandemic of the nineteenth century began in China in 1855 and first spread to India, where it killed over 10 million people. It would eventually spread across the world with famous outbreaks in Hong Kong, San Francisco, Mecca, Glasgow, and Cuba. It was during this pandemic that scientists were first able to identify that a certain kind of bacterium causes the plague. It has the official name *Yersinia pestis*, after its discoverer, Alexandre Yersin. Black rats were discovered to be the hosts of this bacterium, and fleas from these rats infected humans with the disease. These discoveries led to public health measures aimed at reducing rat infestation and the development of insecticides to control fleas. Combined with the subsequent development of a vaccine, the plague gradually became a thing of the past in most countries, and now only small outbreaks resulting in a few thousand cases are reported each year.

World War I saw one of the largest mobilizations of people around the world in history, with over 70 million troops involved. By 1917, many of these troops were mired in horrific conditions, sheltering in crowded trenches in northern France. This would prove to be a perfect breeding ground for a new pandemic that would kill 20–50 million people, many more than were killed in four years of war (about 15 million, including 6 million civilian casualties). The huge loss of life due to the pandemic that began in early 1918 almost certainly contributed to ending the war on November 11, a day now remembered as Armistice Day, or in the United States as Veterans Day. The 1918 influenza pandemic was caused by a virus from the same family as the ones that cause the seasonal flu. Influenza viruses also caused pandemics in 1957, 1968, and 2009.

In addition to the flu pandemics, in the twentieth century, pandemics and epidemics caused by cholera, smallpox, tuberculosis, measles, leprosy, malaria, and human immunodeficiency virus (HIV) resulted in a staggeringly large number of deaths around the globe. Because of dramatic improvements in sanitation and modern medicine (vaccines, antibiotics, and anti-viral drugs) the devastation wrought by pandemics was largely forgotten by us, the inhabitants of the twenty-first century.

But our perpetual battle with infectious disease-causing organisms and devastating pandemics is an integral part of the human story, and this war is not over. The enormous human and economic toll exacted by the COVID-19 disease vividly reminds us that infectious disease pandemics are one of the greatest existential threats to humanity. Most of us lack an understanding of the conceptual frameworks and facts necessary to critically consider issues that have suddenly become pertinent. The goal of this book is to provide readers with a description of how viruses function and emerge to cause pandemics; how our immune system combats them; how diagnostic tests, vaccines, and antiviral therapies work; and how these concepts are a foundation for our public health policies. The narratives interweave scientific principles and ongoing efforts to combat COVID-19 with stories of the people behind the science. After reading this short book, we hope you will be an informed participant in debates about what we can do to safeguard the future by creating a more pandemic-resilient world. As we note in the concluding chapter, it does not have to be the way it has been with COVID-19 again. With the right investments and scientific advances, we could have the knowledge and tools that will help prevent such devastation in the future.

1 Conquering a Pandemic That Raged from Antiquity to the Eighteenth Century

In the history of the war between pathogens and humans, there are two eras. In the first era, the subject of this chapter, humans were suffering from the misery of infectious disease but didn't know what caused them or how to mitigate their effects. Our ancestors observed, however, that individuals who recovered from these diseases did not get the same disease again, and thus could serve as caregivers during an outbreak. We will begin by briefly describing the earliest ideas regarding the origins of this observation, which we now understand to be acquired immunity. The attempt to harness this observation for human good is the history of vaccines. We will describe how the first era of our eternal battle with infectious diseases ended with one of the major achievements of medicine, the vaccine against smallpox. We will tell the tale here of how this procedure, which ultimately eradicated the scourge of smallpox from the planet, was developed slowly by several cultures on different continents in an empirical way without any understanding of how or why it worked. The following chapters will describe the second era when we learned about the origins of infectious diseases and how to combat them.

The Earliest Ideas

Spirits, Demons, and Divine Punishment

We are an inherently curious species. So, the observation that once someone recovers from a disease like smallpox they are not afflicted again led to ideas attempting to explain what was going on. In ancient civilizations (e.g., India, Egypt, and Mesopotamia) diseases were considered to be a punishment from God for sins that a person had committed. In India, upon being afflicted with smallpox, people prayed to a specific incarnation of God called Sitala. This custom is still followed in India when someone in the family has chicken pox or measles, presumably because these diseases also result in skin rashes and pustules that look like liquid-filled raised bumps on the skin.

Not surprisingly, it was speculated in ancient times that individuals who recovered from a serious illness had sinned less than those who succumbed. The reason why one was protected from a disease after recovering from it was because the sufferer had been sufficiently punished. The reason for subsequently being afflicted with other diseases was presumably that the person must have gone on to commit other sins.

Spirits, demons, and celestial objects were also considered to cause diseases. The name "influenza" is derived from the Italian word for influence because the disease was thought to be caused by the influence of stars. Similarly, in Europe, syphilis was supposed to have been caused by an evil conjugation of planets. One implication of the idea that spirits and demons caused diseases was that one could be protected from them by practicing the right rituals. One of the authors (AC) had firsthand experience with such practices. After AC's infant sister died of pneumonia in 1968, his educated but distraught mother visited a "Man of God." This man provided her with an amulet that AC was made to wear to protect him from disease-causing spirits. AC does not remember when in college he stopped wearing it or whether he was admonished by his mother for this action. He has not been struck with terrible illnesses since he stopped wearing the amulet!

Spirits, demons, celestial objects, and divine punishment are not the causes of infectious diseases. The appeal of these earliest ideas regarding the origins of disease and acquired immunity is their simplicity and the peace of mind that comes from taking action to prevent disease. This is why similar ideas persist to present times in some segments of society.

Expulsion or Depletion

Physicians had also started thinking about why people were not usually infected by smallpox after recovering from illness. One type of explanation is the expulsion theory. An example is one attributed to the great Persian physician and scientist Abu Bekr Mohammed ibn Zakariya al-Razi, who lived in the ninth and tenth centuries AD. His name, al-Razi, means that he came from a city, Rey, near Tehran. During his lifetime, he achieved great fame as a physician with his evidence-based approach to examining disease and evaluating various therapies. He was the first to distinguish between smallpox and another skin disease, measles. Besides his many seminal contributions to medicine, he was also a scholar of grammar and other diverse fields. His work in medicine was influential throughout the Islamic world and beyond.

In al-Razi's view, smallpox afflicted young people because they had too much moisture and this led to fermentation of blood. The products of this fermentation process caused pustules to form, and when they burst the liquid that was expelled was the excess moisture. Given this explanation, it was also clear why smallpox did not recur; people who recovered no longer had excess moisture.

Many other explanations of a similar vein were proposed over the years. Girolamo Fracastoro (1478–1553), proposed that seeds (or *seminaria*) that arose spontaneously in a person, earth, or water were the cause of diseases like smallpox. The disease was transmitted from one person to another by transfer of seeds. To explain acquired immunity to smallpox, he posited that all humans were "contaminated" with menstrual blood during birth. When seeds emerged in a person, it caused this contaminant to putrefy or decay. After recovery, one could not be afflicted by smallpox again because the putrefied menstrual blood was expelled in the smallpox pustules. Variants of this theory included ones where, instead of menstrual blood, other fluids from the birth process were considered to be the contaminants that putrefied and were expelled upon infection with smallpox *seminaria*. It is interesting to speculate on why the birth process had such a powerful influence on the worldview of some early physicians.

Girolamo Mercuriale (1530–1606), a contemporary of Fracastoro, pointed out several problems with the expulsion theory. If the need for expulsion of menstrual blood contamination was the cause for smallpox, why were only humans, and not other mammals, afflicted with smallpox? Why did smallpox not exist in Indigenous populations before the Europeans brought it to the Americas? Since the menstrual blood contaminant had been expelled by smallpox, why was there not protection from all diseases after recovery from smallpox? Mercuriale was logically testing the expulsion model to see if it made sense and whether it was consistent with all the data, and found it wanting.

By the way, this is a good example of how science progresses. A model is proposed to explain certain observations. The model is then found to be inadequate for describing new observations. The model is then modified to accommodate the new data, leading to newer predictions. This is exactly what we witnessed in real time when epidemiological models continually revised estimates of deaths due to COVID-19 as new data became available and lockdown measures started to take effect.

New models that were proposed in the seventeenth and eighteenth centuries to account for observations not explained by the expulsion theory continued to rely on disease-causing seeds, but proposed a different function for the seeds. This class of models was called depletion theories. One version of this type of theory, proposed by the English physician, Thomas Fuller (1654–1734), posited that individuals were born with various kinds of "ovula" (like seeds) and that each corresponded to a particular disease. When the appropriate seed was germinated, it gave rise to a specific disease. Upon recovery from the disease, this seed was depleted, and one could not subsequently be afflicted by the same disease. But the seeds for other diseases remained. This model builds on the seed idea and now easily explains why we are protected only from diseases from which we have recovered.

As new observations became available, the depletion idea also became untenable. When a scientific model cannot be modified further to account for new information, there is a crisis because the old model has to be completely abandoned. Then, in what the philosopher Thomas Kuhn referred to as a paradigm shift, a new model is put forth. In later chapters, we will describe the paradigm shifts that led to our modern understanding of infectious diseases and immunity. For now, let us turn to the history of the devastation caused by smallpox, and how it was vanquished by human ingenuity and informed public policy.

A Brief History of Smallpox

From Antiquity to the Early Eighteenth Century

Smallpox is a viral infection that was not only lethal but also horrifying for the patient. It usually began with a high fever, malaise, and muscle and headaches. That would last for 3–4 days. Then a rash and pustules would develop in the mouth, tongue, and throat. Over the next 24 hours, the rash would then progress to cover other parts of the body. Over a few days, the pustules would increase in size and eventually burst. During this second phase, about 30 percent of those infected would die. Those who survived were often left with a disfigured appearance because of severe scarring of the skin and a loss of skin pigment. Their appearance was proof that they had recovered from the disease and would not be afflicted by smallpox again. Thus, they could safely care for the sick. Fortunately, the COVID-19 disease does not leave visible scars, but this means that scientific tests are required to prove that one has been exposed to the disease and is protected for some time.

It is believed that smallpox started to afflict humans in about 10,000 BC. Concrete evidence for its existence in humans 3,000 years ago is provided by Egyptian mummies with signs of the disease. A medical book written in roughly AD 400 in India describes the classic symptoms of the disease, the appearance of pustules on the skin of patients. The first description of smallpox in Europe is in the writings of Bishop Gregory of Tours around AD 580. With the start of European exploration in the fourteenth century, smallpox was spread to Africa and Asia. By the seventeenth century, smallpox was among the top two killer diseases in Europe. In 1849, 13 percent of all deaths in the Indian city of

Calcutta (where one of us was born) were due to smallpox, and between 1868 and 1907 over 4 million people in India died of the disease. Globally, it is estimated that, just in the twentieth century, smallpox caused 300 to 500 million deaths, with roughly 50 million people afflicted every year as recently as the 1950s. These numbers indicate a level of devastation greater than the enormous damage already caused by the COVID-19 pandemic.

Smallpox was brought to the Americas by the Spanish in about 1520 with devastating consequences. It is estimated that between 60 and 90 percent of the population of the New World was decimated by the virus. The extremely high mortality rate in the Americas was probably due to the simultaneous introduction of other highly infectious viruses like measles that the native population had never seen. Smallpox also influenced the outcome of wars. Infected with smallpox, the Aztec armies had no chance against the invading armies of Cortes.

Smallpox epidemics occurred regularly in American cities. Between 1636 and 1698, Boston had six epidemics. The Boston epidemic in 1721 was so severe that many fled to other colonies, thus spreading the epidemic. The epidemics in Boston and New York in the early twentieth century (1900–1903) led to the establishment of government quarantine facilities and mandatory vaccination programs.

The devastating smallpox and plague epidemics led our ancestors to make an important observation. In 430 BC, during the plague of Athens, the Greek historian Thucydides noted that those who recovered from the disease could become caregivers to the sick without consequence. Throughout history, there are many other references to individuals being protected or exempt from a disease after recovering from it. People who had recovered began to be referred to as "immune," derived from the Latin word *immunis* for "exempt."

The observation of immunity inspired our ancestors to devise procedures to try and protect the healthy from smallpox. The Chinese began practicing a procedure to protect people from smallpox as early as AD 1500. The procedure involved collecting scabs from individuals who had a mild form of the disease. The scabs were then converted to a powder. About a month later, it was administered by inhaling powder nasally through a silver tube, left nostril for females and right for males. A week after this procedure pustules formed in the mouth and skin. It was hoped that these symptoms would not be as severe as the full-blown disease or cause death, but this was not always the outcome. Importantly, people who had successfully undergone the procedure did not get the disease when epidemics occurred.

Europeans encountered a similar procedure being used in India in the seventeenth century. Here, the fluid from smallpox sores and scabs was collected, stored for a while, and ultimately mixed with cooked rice to form a paste. Several punctures were made in the skin (arm or forehead) of a healthy person with a needle. These needle punctures were then covered with the paste made with rice. This method spread from India to other parts of Asia and the Balkans.

The methods developed by the Chinese and Indians came to be called variolation. The story of how variolation was brought to Europe is interesting, and many detailed accounts are available. Our story begins by noting the establishment of the British Royal Society in 1660. It is the oldest scientific academy in continuous existence. Because the Royal Society was a prominent entity, many scientists would send their important observations to the Fellows of this society. In 1700, the Royal Society received two letters from British subjects in China describing the Chinese variolation procedure. Neither communication led to any interest among the Fellows, perhaps because of the perceived risk in infecting an otherwise healthy person with diseased material. In December 1713, similar information about variolation was received in a letter from Emmanuel Timoni, a physician who practiced in Constantinople and whose patients included the British ambassadors there. Unlike in 1700, this letter did elicit interest in the variolation procedure.

Enter Mary Wortley Montagu. She was the wife of Edward Wortley Montagu, the British Ambassador in Constantinople. Her brother died of smallpox, and she herself was badly scarred by the disease. Mary Montagu witnessed variolation in Constantinople and given her personal encounters with smallpox developed a keen interest in the procedure. She decided to have her young son variolated by Dr. Charles Maitland, who was serving the British embassy in Constantinople. When she returned to England, she tried to champion the procedure. In 1721, a major smallpox epidemic broke out in London. Fearing for the life of her younger daughter, Montagu requested Maitland to carry out the procedure on her daughter. Maitland carried out the procedure. Members of the Royal College of Physicians were there to observe the procedure and its successful aftermath.

During the 1721 smallpox epidemic in England, a child of Caroline, the Princess of Wales, fell ill. Although it turned out that the child did not have smallpox, the princess got interested in variolation. While there is controversy as to who played the most prominent role, Montagu, Maitland, and Hans Sloane, Bart (the president of the Royal College of Physicians) all played important roles in promoting variolation among the royal family in England. Thus began what was likely the first "clinical trial."

The trial was to be carried out with six condemned prisoners in Newgate Prison in England. In exchange for participating in the trial, the prisoners were to be pardoned afterward. After being variolated in August 1721, five of the prisoners got symptoms of the disease but recovered. The sixth did not get symptoms, and it was then learned that this individual had recovered from smallpox the previous year. One of the women who had been successfully variolated was then sent to care for a child who was afflicted with smallpox. In spite of living in close quarters and tending to the patient, the variolated woman did not fall sick, further validating the efficacy of the procedure. All six prisoners were freed.

Despite the success of the variolation trial with prisoners, Princess Caroline still wondered whether the procedure was safe for her children. To further convince herself that the procedure was safe, the Princess then sponsored the variolation of five orphan children. Variolation did not harm these children. In April 1722, Princess Caroline, now convinced about the safety of the procedure, had two of her daughters variolated. This popularized the procedure among the upper classes of British society. Variolation was tried in Boston around the same time as the clinical trials in England. Benjamin Franklin became a strong proponent of the procedure.

The events of 1721 and 1722 were covered extensively by the press, which helped convince the public that variolation was a safe procedure. It is impossible to overstate the importance of clinical trials for establishing the safety and efficacy of any new vaccine even today. As we will describe in chapter 7, the time required to carry out proper clinical trials is one of many reasons why developing and deploying a new vaccine takes so long. Viewed from our modern perspective of ethics and morality, selecting prisoners and orphans as the subjects of the clinical trials seems highly unethical. Now, of course, participants in clinical trials of vaccines are healthy volunteers, who participate with full knowledge of the risks.

Variolation was practiced in a way that was traumatic for the person being inoculated. The individual was bled and given very little to eat before the procedure was carried out. Variolation involved the preparation of material taken from a person who was infected. We now know that this material contained the smallpox virus. So, even if the inoculum was prepared by an experienced practitioner, serious or lethal illness could ensue. Variolation also resulted in localized outbreaks of smallpox on occasion. Variolated people were therefore housed together after the procedure to prevent spread of the disease. The danger inherent in the variolation procedure and the trauma to the person being inoculated were reasons why variolation was not widely practiced, and most people remained unprotected from smallpox infections. All of this changed with the work of Edward Jenner, who incidentally had been variolated as a child.

Edward Jenner's Paradigm Shift

Jenner was born on May 17, 1749, in the Gloucestershire town of Berkeley in England. After completing his early education and apprenticeship to an apothecary, Jenner became an early student of John Hunter, a Scottish surgeon and doctor. Hunter and Jenner shared many interests, including studying species that hibernated and the migration of birds.

After completing his work with Hunter, Jenner returned home to Berkeley in 1773. It is believed that a milkmaid told Jenner that she did not respond to variolation because she once had cowpox, a relatively harmless disease in cows and humans.

He also heard about this phenomenon from John Fewster (1738–1824), a medical colleague in Gloucestershire who noticed that a young man who had previously been infected with cowpox did not react to variolation. Jenner began to study the connection between cowpox infection and protection from smallpox. These studies took time because cowpox outbreaks in dairies and farms were infrequent. Jenner's studies led him to believe that the origin of cowpox was a disease called "grease heel," which caused inflammation in the skin of horses, and he thought it was transmitted to cows by farmworkers who tended both horses and cows. In cows, the disease affected the nipples and it was passed to dairymaids while milking the cows. It is legend that dairy maids have fine complexions that were envied by noblewomen. The basis of this legend may be that because their exposure to cowpox left them immune to smallpox their faces were not pockmarked.

With his mentor Hunter's encouragement, Jenner did an experiment to test whether cowpox inoculation could protect humans from smallpox. On May 14, 1796, Jenner used the variolation procedure to inoculate a boy called James Phipps with pus from a sore of a cowpox-infected milkmaid named Sarah Nelmes. Because this event was so momentous in medical history, we also know that the name of the cow that infected Nelmes was "Blossom." Two months later, Jenner repeated the variolation procedure on Phipps, but now he used smallpox. When no symptoms appeared, his hypothesis that cowpox inoculation could protect a person from smallpox was validated. He would repeat the procedure with a small number of individuals two years later.

Jenner's experiment represented a paradigm shift in the human endeavor to protect people from infectious diseases. His new paradigm was that one could protect an individual from a deadly illness by inoculating with a material derived from a relatively harmless (for humans) related disease. Thus, unlike variolation, the procedure was largely safe for healthy people.

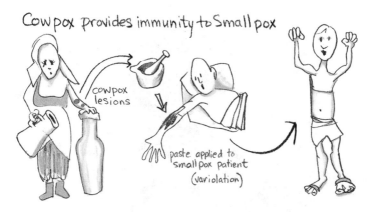

Jenner described his work to the British Royal Society, and he tried to publish a paper based on his findings in the society's prestigious journal, *Philosophical Transactions*, which still exists today. In spite of the fact that Jenner was a Fellow of the Royal Society, the Royal Society rejected his paper. They felt that Jenner did not have enough evidence for his claims, and he might do irreparable damage to his reputation by publishing the work before he had definitive proof. Jenner had his findings published by the private firm of Sampson Low. Presumably, he made money from the sales of this book, which would not have happened if he had published his findings in the *Philosophical Transactions*.

The popularity of Jenner's procedure began to grow. Richard Dunning, an early supporter of the procedure in England, proposed that the procedure be called vaccination because the Latin

name of cowpox is *Vaccinia*. Pasteur later promulgated the use of this term for an inoculation procedure that protected humans from any disease. Ultimately, vaccination replaced variolation, and the latter procedure was made illegal in 1840 in England. Vaccination was made mandatory shortly thereafter.

Jenner's development of vaccination based on careful observation was a remarkable advance. It led to a safe procedure that protected millions of people from a disease that had caused frequent devastating epidemics over millennia. This huge advance in public health was recognized in Jenner's lifetime with numerous honors from professional organizations in England and around the world. The French Emperor Napoleon was supposedly a big fan. Statues were built and poems were written to honor Jenner's contribution to humankind. Jenner's home is now a museum, as is the hut where Phipps was vaccinated. Remarkably, when Jenner did his work, we did not know that infectious microbes cause diseases or that we have an immune system. We will describe infectious microbes and how our immune system works in the next three chapters.

Eradication of Smallpox from the Planet

Throughout the twentieth century, Jenner's smallpox vaccine was administered just like the variolation procedure. After applying the vaccine material, the surface of the skin was punctured repeatedly with a needle to induce a scab. The use of the smallpox vaccine spread around the globe with particularly strong efforts in the Western world. By the early twentieth century, smallpox had been eradicated in Northern Europe and only a small number of cases were reported in other European countries. In 1950, a group of health officials proposed an effort to eradicate smallpox in the Americas and largely succeeded within the decade. In 1958, the Soviet Union proposed that the World Health Organization (WHO) lead an international effort to eradicate smallpox worldwide. Efforts were mounted in almost every country using a novel strategy called ring vaccination. Following the identification of an infected person, every person who lived in the vicinity was vaccinated.

The last two countries with extensive caseloads were Ethiopia and Somalia. A stepped-up focus on these two countries finally succeeded in the eradication of smallpox from these countries in 1979. The last case of smallpox was the fatal infection of Janet Parker, a medical photographer, who contracted the virus from a laboratory that was doing research on smallpox. This resulted in the destruction of all known stocks of the virus except for two vials, one stored in the United States and the other in Russia. As the years passed by without any new smallpox cases, a raging controversy stormed through the scientific community about whether these last vials should be destroyed to literally eradicate smallpox completely from the planet. There was worry that if terrorists were to take control of these stocks and weaponize smallpox, without anyone in the world being immune, a devastating pandemic could take place. Others argued that for scientific purposes it might be necessary to draw on these stocks for some unknown future problem. This controversy was made moot in 2017, when researchers showed that it would be possible to recreate smallpox in the lab using existing methods.

What was the secret to eradicating smallpox, this centuries-old enemy of the human race? The first was the international cooperation of all countries of the world in understanding the horrific nature of smallpox infection and the importance of eradicating it from the earth. Can we now achieve a similar level of global coordination to create a more pandemic-resilient world? The other key was that the virus that causes smallpox did not infect animals. It infected only humans, and spread solely by human-to-human contact. As we will learn in a later chapter, the natural hosts for many viruses are animals, and that viruses can jump to humans when they change in a way that allows them to infect and/or reproduce in human cells. This is precisely what happened to cause the H1N1 influenza pandemic in 2009 and the COVID-19 pandemic. Eradication of a viral disease that infects animals would require the extermination of entire species of animals or finding a way to vaccinate them. Newer technologies have suggested approaches where genetically engineered insects or animals are released into the wild. These organisms are engineered to have the ability to breed with existing species and block the ability of specific pathogenic microbes to survive in their progeny. Whether this is a strategy that is worth trying or whether this may generate unexpected ecological changes is a difficult ethical issue with potentially dangerous environmental impact.

Early Opposition to Vaccination

The smallpox epidemics in Boston in the early twentieth century led the local board of health to start free vaccination programs. Much like flu shots today, one could get vaccinated for free in various locations around the city. In 1902, vaccination was made mandatory in Boston. Those who refused to be vaccinated were subject to a $5 fine or 15 days in prison. Henning Jacobson, a Swedish immigrant in Boston, refused to be vaccinated because he feared that it would make him sick. But, instead of paying the fine, he sued the state of Massachusetts on the basis that the mandatory vaccination program violated his rights. This case went all the way to the US Supreme Court, which ruled in favor of Massachusetts in 1905. The Court's reasoning was that Jacobson's refusal to be vaccinated endangered the health of others.

With improvements in vaccine quality, vaccination is now practiced widely. It is fair to say that vaccination has saved more lives than any other medical procedure. The sharp decrease in child mortality over the past century is primarily due to the success of vaccination programs. Polio, another disease caused by a virus, afflicted 60,000 people in the United States in 1952. Jonas Salk, and then Albert Sabin, developed vaccines that protected people from polio. Today, polio has been all but eradicated from the world.

But, the controversy over the use of vaccines still rages today in some quarters. As we will discuss later, when a significant fraction of the population is not vaccinated against specific diseases, outbreaks occur. Several parents choosing not to vaccinate their children against measles led to the recent outbreaks of this disease in California. High vaccination rates protect the public—in particular, vulnerable people like the elderly and the immunocompromised—by generating something called "herd immunity," which we hear about so much during the COVID-19 pandemic. We will focus on these topics in later chapters.

2 Discovery of Infectious Disease-Causing Microbes and the Dawn of the Modern Era of Vaccines

The ancient Greek physician Hippocrates (460–370 BC) noted that miasmas and poisons in the environment caused diseases. But this was not a concrete concept that could suggest ways to protect humans from specific infectious diseases. In this chapter, we will describe how it became clear that microscopic organisms, or microbes, cause infectious diseases, and how this knowledge led to the development of vaccines against cholera, anthrax, and other diseases in the nineteenth century.

Most of the work in the nineteenth century was done on microbes called bacteria, which are tiny organisms that are made up of a single cell. Bacteria cause devastating illnesses, such as tuberculosis, tetanus, typhoid, diphtheria, and syphilis. These infectious diseases still occur throughout the world. Viruses, which cause illnesses like COVID-19, influenza, and polio, were too small to be seen with the technology available during the nineteenth century. Thus, they were not characterized until the beginning of the twentieth century. We will focus almost exclusively on viruses starting with the next chapter. The knowledge gained by Koch, Pasteur, and others in the nineteenth century about disease-causing microbes and vaccines, described in this chapter, is important background for the chapters that follow.

"Animalcules" under a Microscope

Lenses were used in ancient times to convert sunlight into fire and later on to help with reading (reading stones). But, it was only in the seventeenth century that they were used for scientific exploration. Galileo constructed a telescope in 1609 with which he observed celestial objects, thus creating modern observational astronomy. Although the Assyrians seem to have developed a rudimentary form of a microscope as early as 700 BC, it was not until the seventeenth century that van Leeuwenhoek used such an instrument to reveal the fascinating world of microscopic organisms.

Antonie Philips van Leeuwenhoek (1632–1723) was an amateur Dutch inventor. He was a businessman and held important municipal positions in the city of Delft. Van Leeuwenhoek invented a new way to make lenses that he kept a trade secret. This advance enabled him to construct a microscope that was more powerful than others available at the time. His microscopes were delicately constructed and were tiny, measuring about two inches in length. Van Leeuwenhoek used his new microscope to describe details of the stinger of a bee, the shape of a louse, and mold growing on bread. At the urging of a physician friend, he started communicating his observations to the British Royal Society in letters with hand-drawn illustrations. Later, noting that the water in a nearby pond would become cloudy in the summer, he used his microscope to investigate. He not only saw what we now know as algae, but noticed unusually small organisms swimming about. He named them "animalcules." From his drawings we can discern that van Leeuwenhoek was describing single-celled living organisms that would later be found to be the cause of diseases such as giardia. Observing bacteria for the first time, he estimated their size relative to a grain of sand. He found that more than 100 lined up to equal the length of a sand grain, and therefore, millions could be present in a single drop of water. The Royal Society did not initially believe these observations about tiny living organisms. In 1677, a number of experts were sent to visit van Leeuwenhoek, and he was able to convince them that his observations were valid. Thus, at the dawn of the eighteenth century, a new world of microscopic living organisms was discovered.

van Leeuwenhoek's microscope

But van Leeuwenhoek and his contemporaries did not realize that these tiny microorganisms can cause human diseases. Although some evidence existed earlier, widespread acceptance of the fact that microbes caused disease, the "germ theory," had to wait until the nineteenth century when two great scientists, Robert Koch and Louis Pasteur, provided incontrovertible evidence. The two scientists and their work are forever intertwined in scientific history. The two had very different personalities and were bitter rivals. Each had a different path to success.

The Koch–Pasteur Rivalry and New Vaccines

Koch's Postulates, Anthrax, Tuberculosis, and Cholera

Robert Koch was born in Germany in 1843. His father was a mining engineer. He taught himself to read by the time he was five years old, and was a brilliant student from a young age. After a brief time studying natural sciences in college, he decided to pursue a career in medicine. Koch held positions as a physician in various capacities in Poland, Berlin, and other places, including service as a doctor during the Franco-Prussian War. Koch also developed a deep interest in basic scientific research. Today, we would consider him a clinician-scientist, someone who tries to understand clinical aspects of diseases using basic scientific principles.

Anthrax is a disease that affects both animals and humans, and was a problem in Koch's time. Koch showed that, for a wide variety of animals, he could transfer disease from one animal to another by transferring blood from the infected animal to the healthy animal. All animals thus infected exhibited the same disease symptoms, and had the same rod-shaped bacteria in their blood. This convinced Koch that this specific bacterium caused anthrax. Koch's work on anthrax was the first to associate a specific microbe with a particular disease.

It was known that healthy cattle got sick if they grazed on fields long after anthrax-infected cattle had grazed there. This was a puzzle because Koch had determined that anthrax bacteria in the blood of infected animals lost their infectivity after a few days. He decided that he would need to watch the bacteria over time and would need to develop methods to grow the bacteria in the lab. Koch developed methods to keep bacteria growing for days. This process is called "growing bacteria in culture"—"culture" refers to the medium in which the bacteria are grown. This method is now used millions of times every day around the world. When a doctor suspects that you have a bacterial infection, a small sample is collected from the suspected site of infection (e.g., a wound) and is sent to the pathology department. If the sample contains bacteria, they grow out in culture and can be identified. The doctor can use such a positive test result to prescribe the right treatment to kill the identified bacteria.

With the technique to culture bacteria in hand, using his careful observational skills, Koch noted that on occasion anthrax bacteria would convert into opaque spheres. He showed that these spheres could be dried and then reconstituted weeks later by immersing them into fluid. He suspected that the bacteria, if converted into the dry spheres, or spores, could remain dormant for years. Indeed, this is the case, and they can cause bacterial infection when ingested by uninfected cattle. Some readers will remember the anthrax scares in the United States right after the September 11, 2001, terrorist attacks when an individual placed anthrax spores into envelopes that were sent to members of the US Congress.

As Koch become more skilled in the identification of disease causing bacteria, his methods became codified into rules known as "Koch's postulates":

1. The microorganism must be present in every instance of the disease.

2. The microorganism must be isolated from a human with the disease and grown in culture.

3. The microorganism grown in culture must cause the same disease upon injection in an animal.

4. Samples from the animal in which disease thus occurs must contain the same organism that was present in the original diseased human.

Koch's postulates

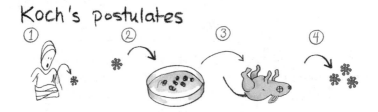

These principles were applied successfully to determine the causative agents of many of the infectious diseases known today. Knowing the identity of specific bacteria that cause a particular disease, scientists and drug companies can develop antibiotics that can kill the bacteria and cure disease. Before the discovery of antibiotics, a small skin cut could get infected and result in death. We live in a world that would be unrecognizable to a nineteenth-century inhabitant because many previously lethal infections and diseases are easily treatable today.

Koch's other significant discoveries were the bacteria that cause tuberculosis and cholera. Tuberculosis (TB) is a disease that has longed plagued the world. It was often called consumption, because it made the person look pale and thin as the disease progressed. In opera, it is the disease from which both Mimi in *La Bohème* and Violetta in *La Traviata* suffer, reflecting a nineteenth-century association of romantic tragedy with this disease. TB caused enormous numbers of deaths in the nineteenth century. Since it is a contagious disease, it flourished partly because of the increased population density in growing cities during the industrial revolution. Throughout the nineteenth century, about one out of a 100 people living in New York City died of tuberculosis, roughly the same percentage as the number of reported COVID-19 deaths in the city and ten times more than die of influenza in an average year.

Until Koch showed that it was an infectious disease caused by bacteria, many thought that TB was an inherited disease. In 1882, using his postulates, Koch identified the causative organism and called it *Mycobacterium tuberculosis*. This discovery led to a better understanding of the disease and the development of TB-specific antibiotics, which along with better sanitation resulted in a significant decline in infections and deaths. However, TB is still widespread and remains a scourge in many parts of the world. In 2018 TB killed 1.5 million people globally. An especially worrisome development has been the recent emergence of antibiotic-resistant forms of *M. tuberculosis*. A vaccine that is used around the world to protect against TB infection has only limited efficacy.

Cholera is a waterborne disease that causes severe diarrhea and vomiting. Cholera outbreaks still cause havoc in the developing world today. The most recent outbreak of cholera was in Sudan in 2019. Another recent cholera epidemic was in Haiti in 2010 following a devastating earthquake. There are indications that, sadly, peacekeepers from the United Nations who came to provide aid may have inadvertently brought the disease to Haiti.

Koch received worldwide fame for his identification of the organism that causes cholera. However, the causative bacterium was, in fact, first described by an Italian physician, Filippo Pacini (1812–1883), many years earlier. During the period from the late 1810s to the early 1860s, there were worldwide cholera pandemics that started in India in the state of Bengal. Pacini was a doctor in Florence, Italy, when the pandemic spread into that city. Using a microscope to examine tissues collected during autopsies of those who had succumbed to cholera, Pacini discovered the bacterium, *Vibrio cholerae*, that causes the disease. Remarkably, few, including Koch, knew of his discovery, perhaps partly because the germ theory of disease was not widely accepted when Pacini described his observations. Better sanitation has made cholera a disease that is nonexistent in the developed world.

Koch, who passed away in 1910, received many significant recognitions for his work, including the 1905 Nobel Prize for Physiology and Medicine. We now turn to the work of his bitter rival, Louis Pasteur.

Pasteur, Rabies, and a New Paradigm for Vaccination

Pasteur was born in 1822 in France. His father was a tanner. Pasteur did not distinguish himself academically as a youngster. After earning a bachelor's degree in philosophy in 1840, he was drawn to the study of science and mathematics. As is true today, in Pasteur's time only the very best students in France were admitted to the École Normale Supérieure. Pasteur was ranked very poorly the first time he took the admission test, but he was ultimately admitted in 1843. This hiccup at an early stage of his scientific career did not prevent Pasteur from going on to make transformative discoveries.

When he was a professor at the University of Strasbourg, in France, Pasteur made a very important fundamental discovery which involved the mathematical concept of chirality. Two similar objects that have non-superimposable mirror images are chiral. The simplest example is our right and left hands—look at images of your hands in a mirror and you will see what we mean. While studying crystals of salts of certain acids, Pasteur demonstrated that molecules can also be chiral, either "right-handed" or "left-handed." He developed a way to detect the handedness of such so-called optical isomers. A good example of handedness is sugar. Sugar is a chiral molecule that is right-handed, and sugar substitutes can be composed of its left-handed optical isomer. The molecule in our body that metabolizes sugar does not act on its left-handed isomer, and thus we do not metabolize it. But our taste buds cannot tell the difference between the right- and left-handed molecules, and so such sugar-substitutes would taste the same to us—a free lunch, so to speak.

Pasteur's next big achievement was inventing a process which was later named pasteurization. One of Pasteur's students was the son of a wine merchant, and he interested Pasteur into thinking about how to prevent wine from spoiling. It was commonly believed at the time that wine spoiled because it spontaneously decomposed into constituents that tasted like vinegar. Pasteur showed that this was not true and that a microbe called yeast was required to carry out these chemical transformations. Pasteur also showed that contamination of wine with various other microbes causes it to spoil. He invented a process to prevent this, which exploited the fact that microbes die at high temperatures. The wine was heated to about 120–140°F, and then sealed and cooled. Although this pasteurization process was invented to prevent wine from spoiling, it is rarely used for this purpose today. Rather, pasteurization is used all over the world to prevent milk from spoiling.

Pasteur also played a significant role in laying to rest the popular idea that many living organisms were spontaneously generated from nonliving matter. As old bread begins to grow mold and maggots suddenly appear in old meat, it wasn't illogical to believe that these changes occurred spontaneously. Evidence against this so-called spontaneous generation theory had already been presented many times by other scientists, but Koch's postulates and an elegant and definitive experiment that Pasteur did in 1859 finally proved to be its death knell. Pasteur stored boiled (pasteurized) water in two curved, swan-necked flasks. Boiling the water ensured that there were no microbes in it when the experiment was started. The construction of the swan-neck flask was such that microbes in the air would get stuck to the walls of the tube and not reach the water if the flask was vertically positioned. Pasteur positioned one flask vertically, and the other was tilted. As time passed, the water in the vertical flask did not show any signs of a developing biofilm (you must have seen such disgusting biofilms when you leave food in the refrigerator too long and microbes grow on it). A biofilm developed in the water in the tilted flask because microbes in the air could reach the water. This demonstration was the end of the spontaneous generation theory.

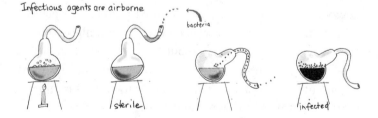

Infectious agents are airborne

bacteria

sterile

infected

Most scientists can only dream of making contributions as important as Pasteur's discovery of optical isomers, his invention of pasteurization, and his experiment ending the debate on the spontaneous generation of microbes. But his contributions to vaccination had such a major impact on humankind that the achievements described above have been completely overshadowed.

Pasteur's paradigm-shifting advance in vaccine development was the result of a serendipitous observation he made while studying chicken cholera. On one occasion, after chickens were injected with the bacteria that causes this disease, they did not fall ill. On further investigation, Pasteur discovered that the batch of chicken cholera he had injected had spoiled. Rather than buy new chickens, he reinjected the first set of chickens with the properly cultured bacteria. To his surprise, the chickens did not fall ill. Pasteur is often credited with the famous remark, "In the field of observation, chance favors the prepared mind." Pasteur's mind was apparently prepared, as he immediately understood that he had stumbled on to an important finding. He realized that you could protect animals from infection with a live disease-causing microbe by vaccinating them with a weakened form of the same microbe.

This was a paradigm shift compared to previous methods. Variolation involved administering the real pathogen. Jenner's use of cowpox involved finding a pathogen that was harmless to humans but related to the one that caused human disease. Pasteur's new method did not involve hunting for a related harmless pathogen or risking the life of the patient by administering the real pathogen. Rather, a weakened or attenuated form of the pathogen could be used. It is worth remarking here that variolation involved powdering material from smallpox scabs and waiting a few days before administering it. These procedures were probably inadvertent ways to attenuate the virulence of the pathogen. But it was Pasteur who in the period between 1879 and 1880 formalized the procedure of using an attenuated pathogen to protect people from infectious diseases, and established a method that continues to be used today. Pasteur labeled his new method of protecting against various infectious diseases "vaccination," in honor of Jenner's use of vaccinia (cowpox) to protect against smallpox. Pasteur used his method to vaccinate birds to prevent cholera and vaccinate sheep to prevent anthrax.

Pasteur then developed a vaccine to protect against rabies. Rabies is an infection of the brain caused by the bite of an infected dog or, more often today, a bat. People infected with rabies exhibit symptoms like paralysis and fear of water. This fear of water is why the disease is sometimes called hydrophobia. Almost everyone afflicted with the disease died. Pasteur was a chemist and not a physician, but having successfully developed two animal vaccines, he was keen to use his skills to cure a human disease or protect people from it. We know today that rabies is caused by a virus, but the concept of a virus was not known at that time. Therefore, Pasteur could neither follow Koch's postulates to identify the causative agent of the disease, nor grow the microbe in culture using methods that worked for bacteria. It was known, however, that the infectious agent was present in saliva. Pasteur is claimed to have been fearless, having used his mouth to suck on a glass tube to draw saliva from a rabid dog.

Using a method developed by his close collaborator, Emile Roux, Pasteur then attenuated the infectious agent. Pasteur and Roux administered the attenuated infectious agent and showed that multiple doses of this vaccine could protect dogs from rabies infection. Pasteur was anxious to try his vaccine in humans. He knew that the onset of symptoms usually lagged the dog bite by about a month. His idea was to vaccinate people soon after the dog bite, and hope that the protective mechanism (about which they knew nothing) would kick in quickly enough to cure them. The mechanism underlying why this might work will become clear to you after you learn about the immune system in chapter 4. The first two patients on whom this procedure was tried were in the late stages of the disease, however, and both died before they could receive the second dose of the vaccine. But Pasteur persevered.

In 1885, Joseph Meister, a 9-year-old boy living in Alsace, was bitten multiple times by a rabid dog that was subsequently shot by the police. His doctor learned that Pasteur had developed a vaccine to treat rabies. In an attempt to evade what was a certain death sentence, he brought Joseph and his family to Paris the next day to seek Pasteur's help. Emile Roux refused to use the vaccine on Joseph as he worried that it was not ready for humans and was too dangerous to try on a child who did not yet have any symptoms of the disease. Pasteur found another physician to administer the treatment and it worked—the boy was cured. Subsequently, others would undergo the same procedure with similar success, and Pasteur became a hero. Years later, Meister, who was devoted to Pasteur, would serve as a caretaker at the Pasteur Institute.

Throughout this period, Pasteur worked on an anthrax vaccine even though Koch, who discovered the bacterium that causes anthrax, was also working on a vaccine. This led to terrible arguments between the two acclaimed scientists. Koch and his students wrote that Pasteur did not even know how to make pure cultures of bacteria. Pasteur fought back. These arguments took on an even more vicious tone during the Franco-Prussian War. In 1868, Pasteur had been awarded an honorary degree by the faculty of Bonn in Germany. He returned it during the war with an angry accompanying note. Thus began a division between German and French immunologists that would continue for decades, to the detriment of scientific advances. Pasteur ultimately achieved success in a public experiment in 1881 when he successfully vaccinated several sheep and cows, and a goat, to protect them from anthrax. He then declared it to be a great French victory. Ironically, an anthrax vaccine had earlier been developed by Jean Joseph Henri Toussaint (1847–1890) in France. Pasteur used the same method as Toussaint, but claimed that his approach was different.

When Pasteur died, he left his laboratory notebooks to his oldest male child, and his will stipulated that these notebooks should never leave the family and were to be passed on from generation to generation by male inheritors. In 1964, Pasteur's last surviving direct male descendant donated his laboratory notebooks to the Bibliotheque Nationale in Paris. Scholars studying these notebooks found that Pasteur often cut corners in his work, sometimes did not describe exactly how experiments were done, and did not always publicly report results transparently. This straddling of ethical boundaries or, worse, fraud is severely punished by the modern scientific community. Indeed, as it should be, because the scientific edifice is built on the trust that scientists have described their studies honestly. Mistakes can happen, of course, but deceit is not allowed.

Pasteur's straddling of ethical boundaries notwithstanding, he made groundbreaking advances that had a transformative effect. Vaccines designed using Pasteur's methods have saved more lives than any other medical procedure. Vaccines that protect children from diseases are a major contributor to the dramatic reduction in childhood mortality. Today, we crave a vaccine against the ongoing COVID-19 pandemic, and hopefully, we will have one soon. Pasteur's work is the foundation for this hope.

For his achievements, Pasteur received many honors and awards. Many streets around the world are named after him, and the Pasteur Institute in Paris is a famed medical research laboratory that Pasteur himself founded. He died in 1895, when he was 72, and his body is interred in the first floor of the original building of the Pasteur Institute. Visitors are welcome to see his tomb and the apartment where Pasteur lived at the end of his life. Pasteur did not receive a Nobel Prize because the first of these was awarded in 1901.

Koch's and Pasteur's work focused mainly on bacterial infections. But there are other organisms that can cause disease, like fungi, parasites, and viruses. Because viruses pose a special threat to humans, causing many recent pandemics, in the next chapter, and in most of what follows, we focus on these tiny organisms.

3 Viruses and the Emergence of Pandemics

When Pasteur was trying to make a vaccine for rabies, he was unable to find the causative agent. Filters were often used to trap and isolate bacteria that might be contained in fluids. These filters were unable to trap the causative agent of rabies. Pasteur reasoned that the microbe that caused rabies must be very small. Rabies is caused by a virus, and indeed it is very tiny. The virus that causes the COVID-19 disease is a member of a family of viruses called coronaviruses. These viruses are roughly spherical in shape and have a diameter of roughly 100 nanometers, which is a thousand times smaller than the diameter of a human hair. The influenza virus, which causes the seasonal flu, is also of similar size. In comparison, the bacterium that causes tuberculosis is rod-shaped, and its length is 20–40 times larger. Pasteur also failed when he attempted to culture the rabies virus using the procedure that Koch developed to grow bacteria. It was only in the twentieth century that methods to visualize and study viruses were developed.

In this chapter, we will describe why viruses need us, and plants and animals, to survive. We will also describe the different types of viruses that exist and how they function, as well as how

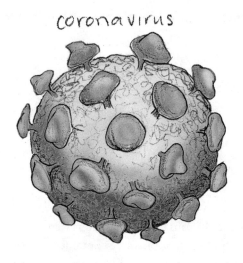

coronavirus

pandemic-causing viruses emerge to wreak havoc. We begin, however, with a bit of history about our long war with viruses.

Our Eternal War with Viruses

Viruses are very simple ancient organisms that have probably existed since life began. For reasons that will become clear in the next section, viruses cannot reproduce on their own. They have to colonize bacteria, plants, and animals (including humans) in order to replicate and propagate their species. Therefore, viruses have specialized skills that let them invade other species and replicate inside them. When a virus invades the human body and replicates, it can damage our cells and tissues. The immune system, about which we will learn in the next chapter, tries to kill viruses that invade us to prevent and combat viral infections. This war between viruses and our immune system has raged since time immemorial.

Pre-agrarian humans contended with fewer types of viruses than we do. These ancient humans were also afflicted by viruses, but they were different in nature from the highly contagious ones that circulate today like those that cause flu, measles, or COVID-19. This is largely because of differences in the lifestyles of modern and ancient humans.

Many viruses that circulate today spread by casual contact between people. Infected people exhibit symptoms of disease and some people die because of the acuity of disease. Our immune system usually succeeds in completely eradicating the virus from our bodies, and we are cured. Remarkably, for a period of time, our immune system "remembers" that a particular virus had previously infected us. If the same virus reinfects us, the immune system can swat it away. For some viruses, this shield of immunity can last as long as a person's lifetime.

Contagious viruses that kill some infected people, but which are normally eradicated from our bodies by the immune system, were unlikely to survive as a species in a pre-agrarian world. Our ancestors at that time lived as small groups of loosely connected people who occupied a large area. So, people encountered very few others in their daily lives. If a virus like that which causes COVID-19 infected a person, that individual would therefore infect very few others. The infected people would either die or recover and be immune to the virus. In a small population, over time, most would become immune, and a newly infected person would be very unlikely to encounter a susceptible individual during the course of disease. Therefore, the virus could not be transmitted to new people in whom it could replicate. Thus, over time, the virus would become extinct. If the virus caused a very lethal disease, it would kill everyone in a small community and become extinct because again there would be no one to infect.

This is why most viruses that circulated in pre-agrarian humans were probably not terribly contagious or deadly, and were not eradicated from the body by the immune system. These viruses adapted to coexist for the lifetime of the infected person, hiding out quietly most of the time. Periodically, they would rear their heads and infect new cells, causing recurrence of disease symptoms. The immune system would then suppress this recurrence, and the cycle would continue. Herpesviruses are an example of such an ancient type of virus that circulates in modern human populations.

Viruses like those that cause the flu, measles, and COVID-19 that were unable to thrive in pre-agrarian times became viable once human ingenuity led our ancestors to learn how to grow crops. In the agrarian society that emerged, people started living together in larger communities concentrated in smaller areas of land. In a dense population, a contagious virus can potentially be transmitted to many others by an infected person. So, the virus can replicate in many people. If the population is large, many individuals have to be infected and then recover to reach a point where a sufficiently large proportion of the population is immune. As this takes a long time, the virus can keep spreading to new people. Furthermore, in a large and dense population, new births provide a constant stream of new susceptible people. This is why highly contagious viruses that cause diseases that our immune system can usually clear from our bodies began to thrive in the agrarian era. Farming also led humans to domesticate animals and live closely with them. Viruses that infected animals and could also replicate in humans began to spread in the human population. Thus, it came to be that a great diversity of viruses began to circulate among humans.

The exchange of individuals between communities increased as travel became easier. These "immigrants" could bring diseases caused by viruses in their communities to others. For example, as mentioned in chapter 1, European immigrants brought a disease caused by a virus, smallpox, to the Americas. Immigration is also a source of new susceptible people for a virus that already exists in a community. When the industrial revolution began, people started to live in cities with even higher population densities than farming communities. Highly contagious viruses flourished even more. The human race is connected today by our shared history of battling the same contagious viruses.

The transition from hunter-gatherer societies to agrarian ones, the industrial revolution, and the many technologies and innovations that followed have improved the quality of human life as measured by a myriad of metrics. For example, we live longer now and childhood mortality is much lower. But the accompanying changes to the way we live also made us more susceptible to infection by a greater diversity of contagious viruses that can cause acute disease. Yet in spite of how the changes in our way of life have favored viruses, we have been winning the war against them. This is because of human ingenuity. We learned to develop vaccines that protect us from many disease-causing viruses. But vaccines take time to develop. So, anytime a new virus emerges, we remain vulnerable to devastating pandemics. The COVID-19 pandemic, caused by the SARS-CoV-2 virus, is only the most recent example.

Let us now dig into how viruses work, how they replicate, and why they cannot do so without us. But first we need a primer on the basic machinery that enables living organisms to function and replicate.

DNA, RNA, and Proteins

All living organisms try to replicate and propagate their species into the future. Since ancient times, people noticed that children share some traits with their parents, and the origin of heredity was hotly debated. But it was only in the nineteenth century that the Catholic monk and botanist Gregor Mendel's careful studies while breeding peas provided the first rigorous basis for heredity. His work led to the concept of genes, which are inherited from one's parents. But Mendel did not know what a gene really was. The discovery of genes had to wait until 1953 when James Watson and Francis Crick, two young scientists working at the University of Cambridge in the United Kingdom, first described what a gene really is. Informed by the studies of many others, including Rosalind Franklin, they had a flash of insight that has transformed how we think of ourselves as individuals and as a species, and indeed our understanding of all living things. Their discovery of how a molecule called DNA stores all our genetic information and faithfully reproduces it in our progeny also laid the foundation for modern medicine.

A DNA molecule is made up of two long strands, each comprised of four types of units that are connected together. The four types of units are called bases, and are labeled A, T, G, and C. The two strands of DNA wind around each other to form a double helix. This is made possible by the fact that A on one strand pairs only with T on the other strand, and G pairs only with C. Therefore, each DNA strand in the helix has a sequence of bases that is the complement of the sequence of the other. The sequence in which these four types of bases are connected in a DNA molecule encodes information about the organism. This is the information that is passed on to progeny.

Complex organisms, like us, are made up of many cells. Each cell contains a copy of our DNA within an enclosure inside the cell called the nucleus. The cells in an organism need to be continually replenished with new cells. This is accomplished by a process by which one cell divides into two identical daughter cells. During this replication process, the DNA double helix in the original cell is copied into two identical DNA molecules. First, the two strands of DNA are separated. Each original strand now serves as a template for the synthesis of a new complementary strand. This is accomplished by a cellular molecule called DNA polymerase, which joins the right complementary bases one by one to the growing new strand. The structure of DNA provides a mechanism to "proofread" the growing complementary strand of DNA. If the wrong base is added, it will not pair with the template strand, and so it is excised and the correct base is then added. In the end, we have the two old strands of DNA, each paired with a new complementary one. Each of the two new DNA helices becomes the DNA molecule in each daughter cell. Of course, errors do occur sometimes, and the errors are called mutations. The error rate for copying DNA in higher organisms

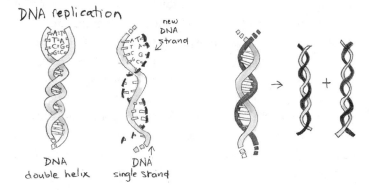

DNA replication

DNA
double helix

DNA
single strand

new
DNA
strand

is small—during each replication cycle, the probability that an erroneous base will be inserted into the new growing strand is roughly one in a billion.

Our cells work together to allow us to perform all our functions. The functions of a cell are carried out by proteins. Proteins make life work. If we imagine that a cell is like a car, proteins make up all the parts that allow the car to function. Proteins are long strands of units called amino acids. There are 20 types of amino acids. So, an enormous diversity of protein sequences can be generated by connecting these amino acids in different ways. For example, as there are 20 choices for amino acids at each position, a string of just three amino acids could be arranged in $20 \times 20 \times 20 = 8{,}000$ different sequences. Proteins are much longer, with an average length of about 400 amino acids. The number of possible proteins, each with a different sequence, that can be created is therefore immense. The sequence of amino acids in a particular protein determines its function. Proteins with different sequences have different functions. Information about all the proteins we can have in our cells is encoded in our DNA.

Ingenious experiments carried out after the structure of DNA was discovered showed how DNA encodes information about the sequence of amino acids in proteins. Different sequences of three contiguous bases in a DNA molecule correspond to different amino acids. For example, AGC corresponds to one particular amino acid, while GCC corresponds to another, and so on. Given that there are four types of bases, there are $4 \times 4 \times 4 = 64$ combinations of three bases. So, our DNA can encode information on 64 types of amino acids. In reality, there are only 20 amino acids. So, multiple types of three-letter strings of bases correspond to the same amino acid. The sequence of three-letter strings of bases in a stretch of DNA (a gene) corresponds to a

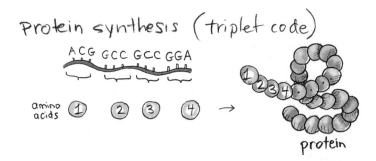

particular amino acid sequence, and hence encodes information on a specific protein. So, you see that DNA, using only a four-letter alphabet, is a compact and ingenious way to encode complex information.

How does the information encoded in DNA get translated into making proteins in a cell? Just as a car needs many bolts to function, cells usually need to make many copies of a protein in order to enable their functions. The gene that encodes information on a particular protein's sequence is first converted into a molecule related to DNA called RNA. RNA is a very ancient molecule, and almost certainly existed before DNA or proteins. An RNA molecule is usually composed of a single strand of connected bases, and looks very similar to a single strand of DNA. The difference is that RNA's four-letter alphabet of bases is not, A, T, G, and C, but A, U, G, and C. So, U replaces T. A cellular molecule, called RNA polymerase, transcribes the DNA sequence that comprises a gene into many RNA molecules that have the complementary sequence. Each RNA molecule now contains the information on the sequence of amino acids in the corresponding protein. A large and complex machine in cells, called the ribosome, then takes each RNA molecule and translates its

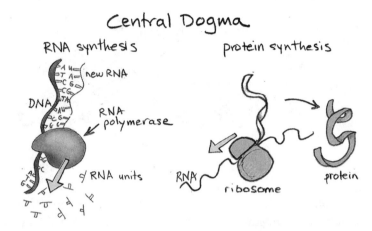

sequence of bases into the sequence of amino acids in the corresponding protein. The way in which the information encoded
in genes in our DNA is first transcribed into RNA and then translated into the synthesis of corresponding proteins is called the
"central dogma of molecular biology."

Viruses Need Us to Replicate

Just like animals and us, viruses have to replicate in order to propagate their species. In fact, their principal function is to make
copies of themselves. To carry out its functions, just like our cells,
a virus needs proteins. Just like our cells, each virus particle also
contains its genetic information. But, unlike each one of our
cells, a virus particle does not have all the machinery needed to
translate its genes to corresponding proteins. Viruses enter a person, animal, or plant and invade their cells. A virus then hijacks
the machinery in the cell they have invaded (e.g., the ribosome)
to translate its own genetic information into many copies of its

proteins. This enables the assembly of many new virus particles that can go on to infect other cells. In a sense, viruses are parasites.

How Viruses Enter Our Cells

Cells respond to changes in their environment in order to carry out their functions. Each type of cell has specific proteins, called receptors, that stick out on their surface to sense the environment. A receptor on a particular cell binds only to a specific substance in the environment. If the receptor binds to this substance, the cell detects its presence and responds accordingly. Viruses have spikes on their surface made up of viral proteins. To cause infection, a virus's spike must bind to receptors on the surface of cells in the tissues that it invades. Once a virus's spike binds to a receptor, it can force its way in through the cell wall and then hijack its machinery to replicate.

The SARS-CoV-2 virus, which causes the COVID-19 disease, binds to a specific human receptor, called ACE2, that helps regulate blood pressure. ACE2 is abundantly present on the surface of cells in the lung. So, an airborne virus can enter through our respiratory tract and bind to ACE2 on our lung cells. This is why SARS-CoV-2 spreads through the air and infects our lungs. But ACE2 is also expressed on cells in the heart, intestine, and kidney, which may explain why other medical problems arise in COVID-19 patients.

Types of Viruses

There are two broad classes of viruses, DNA viruses and RNA viruses. DNA viruses carry their genomic information in the form of DNA, just like we do. The herpesviruses are examples of DNA viruses. RNA viruses carry their genomic information in

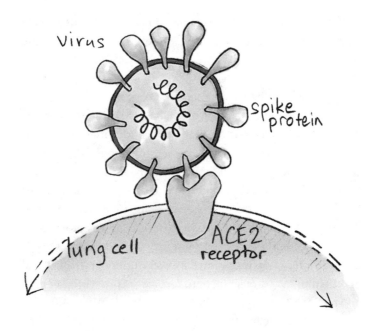

the form of RNA. Viruses that cause common childhood diseases like measles, mumps, and rubella are RNA viruses, as are the ones that cause influenza, AIDS, and the COVID-19 disease. RNA and DNA viruses hijack the host cell's machinery in different ways to replicate.

Once a virus enters a cell, its DNA or RNA genome is released into the cell. The DNA genome of DNA viruses enters the cell's nucleus. Now, new viral proteins are made using the host cell's machinery in the same way that our own DNA is translated into our own proteins. First, some viral genes are transcribed to the corresponding RNA molecules, and then these are translated into some "early" viral proteins. Along with the host cell's machinery, these early proteins help translate the rest of the virus's DNA into a complete set of viral proteins. These proteins

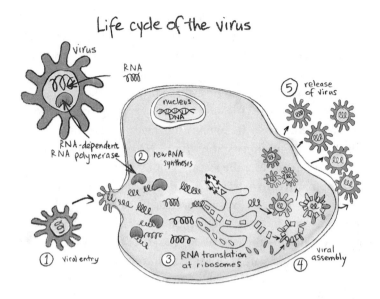

Life cycle of the virus

virus

RNA

nucleus
DNA

⑤ release of virus

RNA-dependent RNA polymerase

② new RNA synthesis

① viral entry

③ RNA translation at ribosomes

④ viral assembly

are then assembled into new virus particles, which then exit the cell and search for new cells to infect. Sometimes, the host cell dies during this process.

Once inside the cell, RNA viruses need to make many copies of their RNA genome, which can then be translated into many viral proteins, which are assembled into new virus particles. Most RNA viruses carry a molecule that makes copies of their own RNA. This molecule is called an RNA-dependent RNA polymerase. Once many copies of the virus's RNA are available, the host cell's protein-making machine, the ribosome, is used to translate the information in the RNA to the virus's proteins. New virus particles are then assembled and they bud out of the cell.

While most RNA viruses replicate as described above, one type of RNA viruses does it differently. These viruses, for example,

HIV, are called retroviruses. Retroviruses first convert their RNA into the corresponding DNA sequence. This is accomplished by a protein that the virus carries with it, called reverse transcriptase. The viral DNA then enters the cell's nucleus, and a viral protein called integrase makes a nick in the host cell's DNA and inserts the virus's DNA there. Now, the host cell's genome is altered forever, as it contains this piece of viral DNA. Using the host cell's machinery, the virus's DNA is then translated into its proteins, enabling the assembly of many new virus particles that go on to infect new cells. Until the mechanism by which retroviruses replicate was discovered, it was believed that the only way that genomic information was translated into proteins was by following the central dogma—DNA to RNA to proteins. Retroviruses translate their genomic information to proteins by the route, RNA to DNA to RNA to protein. For their paradigm-shifting discovery of how retroviruses replicate, David Baltimore and Howard Temin were awarded a Nobel Prize in 1975.

Retroviruses were first identified as tumor-causing viruses in animals. Whether they played any role in human cancer or any other human disease was not clear until the 1980s when a

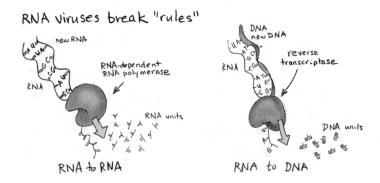

RNA viruses break "rules"

retrovirus was identified as the cause of a rare leukemia. Subsequently, HIV, a retrovirus, was found to be the causative agent of AIDS. When the human genome was first sequenced, a big surprise was that a large percentage of our DNA genome was comprised of retrovirus genes. These genes are usually not translated into proteins. This tells us that we have waged war with retroviruses for a long time, and they are hiding in our DNA. Understanding whether the retrovirus genes in our genome are implicated in cancer and how they impacted evolution of the human species is an active area of research.

RNA Viruses Can Change Guise Rapidly

When our cells translate the information in our DNA genomes into corresponding RNA and then into proteins, very few errors are made. RNA-dependent RNA polymerase, however, makes mistakes at much higher rates when it copies an RNA virus's RNA. Similarly, reverse transcriptase makes many errors when it creates a DNA molecule from a retrovirus's RNA genome. So, the proteins of RNA viruses that are made in a host cell often have a variety of amino acid sequences that are different from the proteins of the virus that originally infected the cell. Most such mutations result in defective proteins that prevent the new virus from functioning. But some mutations allow the virus to function just as well, and some even better. Thus, RNA viruses have an ability to change guise quickly and still function. Sometimes, they can even evolve new functions.

Using Genomic Information to Test for Viral Infection

During an ongoing epidemic or pandemic, it is very important to be able to rapidly identify infected people. These are the people who need to be isolated from others. To detect the virus, we need to know its RNA or DNA sequence. Once we know the sequence, a standard method called polymerase chain reaction (PCR) can be used to detect whether a sample of human fluid contains the virus that is the causative agent of a disease. PCR is a simple, ingenious idea for which its inventor, Kary Mullis, won a Nobel Prize. It takes advantage of a small synthetic DNA fragment called a primer that is designed to bind to the specific viral DNA or RNA sequence. If the RNA or DNA sequence specific to the primer is present, addition of a DNA polymerase allows a double-stranded fragment of DNA to be made. Doing this repeatedly many times amplifies this DNA fragment, and when there are many copies of DNA, it is easy to detect. This enables highly sensitive detection of rare DNA or RNA sequences. A few days after the RNA sequence of the virus that causes COVID-19 was published, Christian Drosten in Germany published a PCR method for detecting this virus. Implementing this method only required ordering a set of primers, and it was rapidly adopted by most countries around the world. However, the Centers for Disease Control and Prevention (CDC) in the United States decided to design its own test, which delayed the introduction of testing in that country.

Examples of RNA Viruses and Why They Cause Pandemics

In this section, we describe three examples of RNA viruses that have caused pandemics in the last century, and how the pandemic-causing viruses emerged.

SARS-CoV-2

Coronaviruses are a family of RNA viruses, and for years, four different types of these viruses have circulated in the human

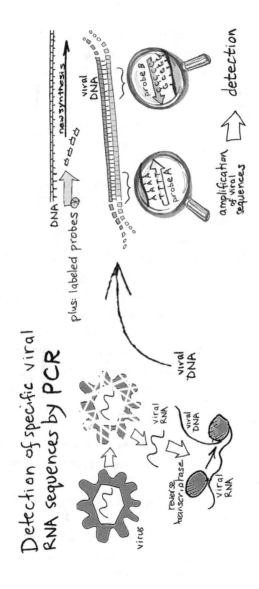

population. Some of them are among the many types of viruses that cause the common cold. No one pays much attention to them because these viruses cause mild disease symptoms in the vast majority of people whom they infect. They are basically just a nuisance.

In 2003, many patients in China were found to be suffering from acute respiratory distress syndrome that was caused by severe lung damage. Soon it was determined that the disease was due to a new type of coronavirus. The virus was called severe acute respiratory syndrome coronavirus, or SARS-CoV. The virus started spreading across East Asia to Hong Kong and Singapore, and then to Canada. SARS-CoV was deadly, resulting in the deaths of about 10 percent of those infected. Strong public health measures in China and other countries were ultimately able to control the SARS epidemic. About a decade later, in 2012, a virus emerged from Saudi Arabia that spread around the world, but mainly to South Korea. It too caused a severe respiratory illness, which was called Middle East respiratory syndrome (MERS). The causative agent was again a coronavirus. This virus was even more lethal than SARS, killing about 35 percent of those that it infected. Again, strict public health measures were able to extinguish MERS.

In late December 2019, physicians in Wuhan, China, began to suspect that a new virus was responsible for a flu-like respiratory disease. On January 10, 2020, Chinese officials announced that a new coronavirus caused this disease, and published its RNA sequence. The sequence was more similar to the virus that caused SARS than the one that caused MERS. In late January, almost on the same day, South Korea and the United States both detected patients who tested positive for the novel coronavirus. Within months, the virus would spread around the world, resulting in the death of hundreds of thousands of people. In response to the fast-spreading pandemic, countries around the world shut down their economies to keep people apart. The cost of this worldwide economic catastrophe is measured in many trillions of dollars, and millions have lost their jobs. The novel coronavirus that devastated the world came to be called SARS-CoV-2, and the disease it causes, coronavirus disease of 2019 (COVID-19).

Why do new viruses cause pandemics or epidemics? How do these new pandemic- or epidemic-causing viruses emerge? Why did SARS and MERS not spread around the world, while COVID-19 became a global pandemic? Let us consider these questions in turn.

When a virus has been circulating in the human population for a while, most people have some level of immunity to it. So, they can fight the virus adequately, and only a few people become ill. When a totally new virus emerges, no one has immunity to this virus. So, the virus can infect anyone, and if the virus is easily transmitted, infected people can spread the virus to many others they encounter. If the virus also causes a lethal disease, a frightening pandemic can result. The coronaviruses that caused SARS, MERS, and COVID-19 were such new viruses to which humans were not immune.

How did these new coronaviruses arise? As we discussed earlier, RNA viruses mutate, and this provides them with a mechanism to change guise and evade human immunity. Did new coronaviruses, like SARS-CoV-2, evolve from mutations in the coronaviruses that circulate in humans and cause mild diseases, like the common cold?

To illustrate a point pertinent to this question, let us consider an RNA genome with 10 genes. Suppose that, due to errors

made by RNA-dependent RNA polymerase, the chance of mutations arising in a gene when it is copied is one in five. On average, there will be mutations in two genes every time this RNA
is replicated. If the RNA genome had 20 genes, there would be
4 mutations, if it had 80 genes, there would be 16 mutations,
and so on. So, an RNA virus with a longer genome should have
mutations in more genes. Compared with other RNA viruses,
coronaviruses have a very long RNA genome. But when we peer
at RNA sequences of these viruses, we do not see many mutations. Indeed, SARS-CoV-2 also has not mutated very much since
it was first identified. Why is this?

Genes encode information about proteins that have to work
together to enable a virus to function. Mutations make a protein
different, which is likely to make it less compatible for working with other proteins. The larger the number of proteins with
mutations, the more difficult it will be for viral proteins to work
together, increasing the chance that the mutant viruses will not
be viable. So, coronaviruses, with their long RNA genomes, would
produce many progeny that would likely not function because
of replication errors made by RNA-dependent RNA polymerase
in many of its proteins. This is why coronaviruses have a protein
that serves as a proofreader, like the one our cells have for DNA
replication. So, if a wrong base is inserted as a new RNA strand is
being created, the proofreader can excise it and the error is cor-
rected. This is why coronaviruses mutate less than expected. The
spikes of most coronaviruses that circulate in humans and cause
mild disease do not bind to ACE2 (the receptor for SARS-CoV
and SARS-CoV-2). They enter cells by binding to a completely
different receptor. In order to bind to ACE2, many new mutations would have to arise in their spike proteins. Thus, since
coronaviruses have a proofreader, it is unlikely that mutations

in the common human coronaviruses led to the emergence of SARS-CoV or SARS-CoV-2.

Some families of viruses not only infect humans but can also infect and propagate in animals. For example, coronaviruses also infect rodents and bats. Although these viruses share some features with the human coronaviruses, their proteins are different in important ways. These differences usually make it difficult for a member of a virus family that thrives in a particular animal to do so in humans. Why this is so is explained by Darwin's theory of how species evolve to adapt to their surroundings. Darwin's studies showed that organisms mutate randomly. Most mutants are likely to be less adapted to their environment than their parent. If by chance a mutation arises that is better adapted to the environment than the parent, it slowly outcompetes the older species and becomes dominant in the population. In an analogous manner, mutations arise over time in coronaviruses that infect bats, for example, and they become better adapted for multiplying in bats. But these changes make the coronaviruses that infect bats less suitable for thriving in humans because we are a different environment.

However, every now and then, changes can occur in a virus that allow it to jump from being a virus that thrives in an animal to one that can productively infect humans. Often, this jump occurs in stages. For example, a bat coronavirus through a few mutations could become capable of infecting another animal. Also, bats can harbor many types of coronaviruses at the same time, and so two different coronaviruses could coexist in an infected cell. The two viruses could swap pieces of their genomes with each other to create a new hybrid virus that can infect another animal or a human. This process wherein the genomes of two viruses mix is called recombination. If the new hybrid bat virus infects an intermediate animal, it could acquire a couple of additional mutations therein that makes it thrive better in humans. This is especially likely to be true if the intermediate animal shares some traits with humans. SARS, MERS, and COVID-19 are diseases that were almost surely caused by viruses that jumped from bats to us, perhaps through intermediate animals.

During the first SARS epidemic, it was first thought that SARS-CoV was passed to humans by civets. These are small, cat-like animals that were being sold at live animal markets in southern China. There were documented examples of humans being infected by civets. But later it became clear that the SARS virus did not originate in civets. This initiated a worldwide search for the origin of the animal from which SARS-CoV jumped to humans. In 2005, working with an international consortium, Shi Zhengli, a Chinese virologist from the Wuhan Institute of Virology, identified coronaviruses in bats in China that were closely related to the SARS virus. This suggested that bats were the original source. Over the next 12 years, Shi traveled throughout China exploring

bat caves. She collected and cataloged the coronaviruses that she found in bats throughout China. In 2017, she finally discovered what she was looking for. In a cave in southern China, she found a bat colony with viruses that were almost identical to the SARS virus, and were therefore its original source. Given her unusual devotion, the popular press dubbed her China's "Bat Woman."

When COVID-19 first emerged in China, Shi was urgently summoned home to Wuhan and she was the first to sequence and analyze the new virus. She reported that the new virus was related to the SARS virus, but, remarkably, it was almost identical to a bat virus that she had collected earlier. The major difference was in the proteins that made up the virus's spike. The spike protein was very similar to that of a virus isolated from a pangolin, a small mammal with an armadillo-like shell. This suggested that the virus had passed from the bat to the pangolin before making the jump to humans. But this picture is uncertain, and will be clarified as more data become available. Whether this virus further adapted to thrive in humans after directly infecting humans, or whether it passed through an intermediate animal like a pangolin, is unclear.

SARS and MERS were lethal viruses, which rapidly killed many of those that it infected. But neither of these viruses caused global pandemics. Humans infected with these viruses quickly felt very sick with cough, fever, and malaise, and sought medical help. A virus that immobilizes infected people early in this way cannot spread too widely. This is because infected people largely come in contact only with close family members and healthcare professionals. So, it is relatively easy to contain the virus by isolating healthcare workers and close family members who came in contact with infected persons. This is how SARS was rather quickly eradicated. In the case of MERS, the virus jumped from bats to camels and then to humans. MERS infections occurred mainly in people with close contact with camels. All of these factors served to limit the spread of these two coronaviruses.

Comparing SARS-CoV and MERS with SARS-CoV-2 (the virus that causes COVID-19) reveals the kinds of features a virus needs to acquire in order to cause a worldwide pandemic. A person infected with SARS-CoV-2 can spread the virus to others before feeling any symptoms. Many infected people have mild or even no symptoms at all. So, infected people can move around and infect many others before realizing that they are infected. This allows the virus to spread rapidly through populations. Although SARS and MERS have a higher fatality rate than COVID-19, the latter is deadly and kills about 1 percent of infected people. Because it infects many people and is quite lethal, SARS-CoV-2 has features that make it almost perfect for causing a deadly pandemic. The saving grace is that it does not mutate much, which would greatly complicate efforts to design a vaccine that can protect us from it (you will have to wait until later chapters to see why).

Influenza

Influenza is an RNA virus that is not a member of the coronavirus family of viruses. We are very familiar with influenza because it causes the seasonal flu, which like COVID-19 is a respiratory illness. Influenza has a relatively low rate of infectivity, but it can also be transmitted a few days before symptoms appear. Also, like COVID-19, a significant fraction of those infected do not feel very ill, which facilitates the spread of infection. Influenza clearly has a seasonal preference because cases peak in the Northern and Southern Hemispheres during their respective winters. Various factors are responsible for the seasonal preference. Some important factors are that people spend more time with each other in close proximity indoors when the weather is colder, and the virus survives better in cold, dry weather than in warm, humid weather.

The protein that makes up the viral spike of the influenza virus has a long name that is usually abbreviated as HA. Another important protein that is displayed on the surface of the virus is a protein whose name is abbreviated as NA. There are 18 different types of HA and 11 different types of NA. The different families of influenza viruses are classified by the specific combination of HA and NA that they have.

Influenza is an RNA virus that, unlike coronaviruses, does not have a proofreader. So, its proteins, including HA and NA, mutate continuously. For reasons that will become clear in the next chapter, humans often mount strong immune responses against influenza that target HA and prevent it from binding to its receptor. This prevents the virus from infecting new cells, and the infection is controlled. Vaccines also elicit immune responses designed to target HA and thus prevent infection. As the winter progresses, many people develop immune responses to the HA proteins in the circulating strains of the influenza virus, due either to natural infection or to vaccination. Thus, they become immune to these viruses. The virus mutates, and the mutant strains that are both functional and able to evade this shield of immunity established in past years then prevail among the circulating strains the following year. Thus, the war between influenza and our immune systems continues every year.

The World Health Organization has a network of people and countries who surveil the world, sequencing RNA from influenza viruses. Based on this data, and which viruses have circulated among humans in past years, a group of experts make educated guesses about which strains of influenza are likely to be prevalent in the next year. Next year's vaccine is then designed. It usually takes several months to manufacture millions of doses of next year's flu shot, and so this decision is made well before the flu season begins. Even if the educated guess that leads to the design of the vaccine is not correct in some years, since the prevailing mutant strains of influenza that circulate in a given year are usually not too different from ones in past years, most people have partial immunity to them. So, illnesses and deaths are mostly confined to vulnerable groups, such as the elderly and the immunocompromised. But influenza nevertheless kills anywhere between 15,000 and 60,000 people every year in the United States.

The number of deaths and hospitalized patients changes dramatically when influenza pandemics arise. There have been four such pandemics in the last century, in 1918, 1957, 1968, and 2009. Twenty to fifty million people died during the 1918 pandemic, when the world population is estimated to be almost four times smaller than today. How do influenza pandemics arise?

The RNA genome of the influenza virus is made up of eight discrete segments, each of which encodes information about one or two of its proteins. Influenza viruses can also infect animals, such as birds and pigs. Of course, these viruses are different because they have adapted to live in their host animals. However, each of the eight segments of the influenza genome is like a cassette that can be taken out and replaced by another variant of the same gene segment. For example, one of the eight segments encodes information about HA, which is the spike protein. A particular variant of this segment could be swapped for another variant. If the new spike protein is compatible with the proteins encoded by the other gene segments, you have a viable virus. But now the virus has a completely new HA spike to which humans have no immunity, and such a novel influenza virus could cause a pandemic. Indeed, influenza pandemics occur when gene segments of the virus circulating in humans are swapped with those of a bird or a pig.

The 1957 pandemic occurred when three gene segments in the influenza virus that was circulating in humans at the time were swapped with those from a bird influenza virus. The swapped segments included the one corresponding to HA. The 1968 pandemic arose when two gene segments of the strain then circulating in humans, including the one for HA, were swapped for those from birds. The bird viruses are not well adapted to thrive in humans, but since the majority of gene segments in the pandemic causing viruses that emerged in 1957 and 1968 were from viruses that were already circulating in humans, they thrived in the human population. Swaps of gene segments from different influenza viruses usually occur during coinfections. A person who works closely with birds that harbor influenza viruses could be infected by a bird virus and a human virus at the same time, and the viruses might swap their gene segments by a process called gene reassortment. These swaps likely do not produce viable viruses most of the time, but sometimes, as in 1957 and 1968, they do so with devastating consequences.

The 2009 pandemic, however, was caused by a virus that emerged in pigs and directly infected humans. This was likely because multiple viruses derived from humans, pigs, and birds were circulating in pigs at that time. It is thought that a number of reassortments occurred within pigs until the pandemic-causing virus that thrived in humans emerged. We are much more similar to pigs than to birds, which may have helped the direct jump of the virus from pigs to humans in 2009.

Human Immunodeficiency Virus (HIV)

In Africa, a relative of the HIV retrovirus circulates in many species of primates like monkeys and apes. It is believed that HIV,
the form of this virus that can infect humans, jumped from a
certain kind of chimpanzee to us. Hunters who captured chim-
panzees for meat (bush meat) came in contact with blood from
these animals and ate the meat. This sort of contact with a large
amount of the animal virus likely allowed a few mutant forms
of the virus that could multiply in humans to emerge in a few
individuals. Using computational approaches based on viral
sequences and rates of mutation, scientists now believe that HIV
started circulating in humans perhaps as far back as the 1920s,
primarily around the Republic of Congo in Africa. It was only
in the early 1980s, however, that a number of unusual cases of
lung and mouth infections and cancer, all among young men in
California and New York, raised the alarm that a new type of disease might be spreading. This is because these conditions usually
arise in people whose immune systems are compromised, and
young healthy people are not normally immunocompromised.
In 1981, Dr. Michael Gottlieb, an immunologist at the University of California Los Angeles Medical Center, published a report
of the cases in California among young gay men in the CDC's
Morbidity and Mortality Weekly Report. This marked the beginning
of the pandemic disease that we call acquired immunodeficiency
syndrome (AIDS).

In 1983, Françoise Barré-Sinoussi and Luc Montagnier at the
Pasteur Institute in Paris announced that they had identified the
virus that is the causative agent of AIDS. In 1984, Dr. Robert
Gallo at the National Institute of Health in the United States,
who had previously discovered the first human retrovirus,
reported that his laboratory had also identified the retrovirus

that caused the disease. Soon it was realized that the two viruses were identical. A test was then developed, which is used to this day. There was a bitter patent rights dispute between the United States and France, which was ultimately resolved in 1987 when Presidents Reagan and Chirac agreed to share the profits, and donate the bulk of it for AIDS-related research and treatments. Barré-Sinoussi and Montagnier shared the Nobel Prize in 2008 for their discovery of HIV.

HIV is transmitted to others by exchange of bodily fluids, such as blood, semen, and milk from lactating mothers. Before tests were available, tainted supplies in blood banks infected hemophiliacs. Initially, it was thought that the disease infects only gay men or intravenous drug users, but this is not true. Heterosexual sex is the principal cause of HIV infections in sub-Saharan Africa, the epicenter of the disease today. To date, HIV has infected almost 75 million people, and as many as 40 million people have likely died from complications associated with AIDS. In South Africa, there are still approximately 1,000 new infections every day.

Upon initial infection, HIV causes flu-like symptoms that then go away. The virus infects a cell that plays a key role in coordinating our immune responses, resulting in a decline in the number of these cells. This is why patients have a compromised immune system, which results in vulnerability to infections that our immune system normally controls with ease. Without treatment, ultimately, the numbers of the immune cells decline to very low levels, resulting in death. Today, innovations in HIV drug treatment (described in chapter 6) keep the virus under control in treated individuals. But the virus is not eradicated in these people, and it comes right back if treatment is interrupted. The search for a cure for HIV is an active area of research, as is the search for a vaccine. After over 30 years of effort and enormous expense, we still do not have a vaccine that can protect against HIV infection. As we will see in chapter 7, this is because HIV mutates at a very high rate. Fortunately, this is not the case for the virus that caused the COVID-19 pandemic.

Why Do RNA Viruses Cause So Many Pandemics?

As we have described, most RNA viruses mutate quite a bit. Also, their genomes are relatively malleable, which makes possible reassortment of genes as occurred during influenza pandemics, or recombination of genes as that which might have resulted in COVID-19. This malleability and the circulation of related forms of RNA viruses in animals make the emergence of new RNA viruses a perpetual existential threat to humanity. For example, an influenza pandemic is always waiting to happen. Various forms of influenza viruses circulate among birds and pigs. Humans interacting closely with pigs and poultry in farms and markets can facilitate exchanges of the influenza virus between them and these animals. The global population will have no immunity to the new virus, and if, like SARS-CoV-2, it is easily transmitted by casual human contact and is quite lethal, a pandemic will result. As the COVID-19 pandemic has made vivid, this threat is not localized to particular nations or peoples.

Examples of DNA Viruses

DNA viruses are a different beast. Most have a double-stranded helical DNA genome, just like us. As we described earlier, double-stranded DNA can be copied with very high fidelity. So, mutations are rare, which allows DNA viruses to have much longer genomes than RNA viruses. For example, the herpesvirus genome has around 80–100 genes, and smallpox has around 200. Most RNA viruses contain about 10 genes or fewer. With many more genes, DNA viruses are more complex machines and do not change guise as rapidly as RNA viruses.

DNA viruses are very familiar to us, especially those that belong to the herpes family. The herpesvirus family includes the viruses that cause chicken pox and shingles (varicella zoster), cold sores and contagious genital rashes, and mononucleosis (Epstein–Barr virus). Another herpesvirus known as cytomegalovirus (CMV) is also prevalent. Most people are infected by one or more herpesviruses by the teenage years. These viruses are passed from human to human in different ways, but close human contact is usually necessary.

Herpes is an ancient virus, which inserts its genome into the nucleus of the host cell. So, after infection, the cell is permanently infected. For most of us, the virus stays silent after one recovers from the initial infection, and never bothers us again. But in some cases, the virus can reawaken and make new virus particles. This is what happens during a shingles outbreak. After exposure to chicken pox in childhood, the virus hibernates in cells of the nervous system. Usually in older adults, the virus can reawaken and cause a reddened, painful, skin condition, which we call shingles. We really do not understand why the virus gets activated again. It is likely that the immune system normally keeps the virus under control. But in stressful circumstances, or when the immune system is suppressed, herpesviruses are reactivated. Reactivation can occur with all herpesviruses, not just the one that causes chicken pox.

With this understanding of the enemy, viruses, let us turn to our immune system, which combats these scourges on the planet.

4 Immunity

The immune system is our department of defense. It serves as a shield that protects us from invasion by infectious, disease-causing microbes, and combats them if they succeed in infecting us. Our immune system also interacts with other physiological systems to influence health and disease in myriad important ways. In this book, we will focus on how the immune system functions to combat, and usually vanquish, viruses. The concepts that we describe are also generally applicable to how our immune system battles other microbial pathogens.

A remarkable feature of our immune system is that upon infection with a virus, we can mount an immune response that is specifically tailored for it. Think about how amazing that is: the SARS-CoV-2 virus, which causes the illness COVID-19, did not exist when most people living today were born, yet the immune system of infected people responded in a way that was specifically designed to combat this virus. Equally remarkably, our immune system remembers past infections. It responds rapidly if the same virus reinfects us and swats it away before it can cause disease. Assuming SARS-CoV-2 is like most viruses that we encounter, people who have recovered from COVID-19 are likely to be protected from reinfection for at least some time. For some viruses, immune memory can protect us for many years, even for life. This virus-specific memory is the basis for vaccination. An effective vaccine elicits memory immune responses that are specific to a virus against which we wish to protect the population.

In this chapter, we will explain how our immune system mounts responses that are tailored to a particular virus, and how memories of past infections are formed to protect us from reinfection. Our narrative roughly follows the historical order in which various aspects of how our immune system functions were discovered. In the last section, we provide a brief summary of how the immune system works.

The Dawn of the Modern Era of Immunology

Blood or Cells

Wounds and microbial infections often lead to inflammation—that is, swelling, redness, heat, pain, and the accumulation of pus. In the nineteenth century, it was known that plants and animals were composed of cells. It was also known that a strong inflammatory response was often a harbinger of a bad outcome for the patient and that pus was composed of dead cells. Therefore, inflammation was considered to be harmful, and the participating cells were considered to promote illness. It was even thought that these cells could spread disease-causing microbes throughout the body. In spite of these prevailing beliefs, in the late nineteenth century Ilya Metchnikov made the heretical proposition that the cells involved in inflammation were the good guys, the body's first responders to infection and injury.

Metchnikov was born in Russia in 1845, but later emigrated to France where he became known as Élie Metchnikoff. He was a zoologist by training, and had a long-standing interest in the digestive processes of animals. This interest may have influenced his thinking when he made an important discovery about how our immune system works.

One day in 1882, when Metchnikoff was working in Messina, a city in Sicily, he was observing starfish larvae. The starfish were transparent, and so he could watch their cells moving around. A sudden thought occurred to him. What if our cells that moved around could travel to the site of a wound or infection, and surround it to isolate and eliminate the problem? Metchnikoff decided to test his idea that night by inserting a splinter into a starfish. To his great satisfaction, he woke up in the morning to find that cells of the starfish had surrounded the splinter. He then thought that these cells might be trying to eliminate the invader by eating it. Thus began Metchnikoff's lifelong work on trying to establish that cells of the immune system homed in on an area of microbial infection and then destroyed the microbe by ingesting and digesting it. The word *phagein* in ancient Greek means "to eat," and so he labeled these cells phagocytes, and the process by which they ingest invading microbes, phagocytosis. Metchnikoff published a paper describing his observations in starfish, and proposed that inflammation was a beneficial response to infection or injury.

Ideas developed over millennia had led to the belief that inflammation was bad and that an imbalance of the humors, like blood, led to disease. Correcting this imbalance was thought to be critical for restoring health. This was why blood-letting was considered a therapy for many diseases. Given this background, Metchnikoff's proposal was met with strong objections. Two camps emerged: the cellularists, who supported Metchnikoff, and the humoralists, who believed that healing was mediated by blood components. Akin to the Koch–Pasteur rivalry, the cellularists were mostly French, from Metchnikoff's adopted country, and the humoralists were mostly German. The humoralists tried to show that blood products could neutralize the effects of various microbes and toxins. The cellularists believed that phagocytic cells were critical for responding to and getting rid of infectious, disease-causing microbes.

In 1890, Emil von Behring and Shibasaburo Kitasato reported important findings that supported the humoralists. The bacterium that causes diphtheria infects the mouth and throat and results in severe damage to these tissues, often causing suffocation and death. It was known that a toxin produced by the bacteria made it so lethal. Von Behring and Kitasato reasoned that if they vaccinated individuals with just the diphtheria toxin instead of the bacteria, they might generate an immune

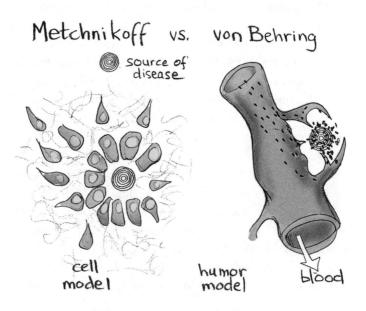

response that would block the toxin. Their experiment worked and they found that after immunization, there was something in the blood that specifically bound to the diphtheria toxin. It was then shown that transferring blood containing this antitoxin could protect the recipient from diphtheria's effects. As tens of thousands of children died in Germany every year from diphtheria, von Behring worked with the Hoechst chemical and pharmaceutical company to develop these discoveries into a working therapy. Interestingly, well over a century after von Behring and Kitasato's discovery, injecting people with blood products from people who have recovered from COVID-19 is being pursued as a therapy. Von Behring was rewarded for the discovery of this form of therapy, called serum therapy, with a Nobel Prize in 1901. Notably, Kitasato did not share this recognition.

Von Behring and Kitasato's discovery was a great victory for the humoralists since a substance contained in blood was shown to be curative. So, although evidence kept trickling in that phagocytic cells played a role in the immune response, the cellularists had lost. Metchnikoff shared the 1908 Nobel Prize for his work on immunology, but it was generally believed by then that his observations, and those of his supporters, were unimportant for understanding immunity. For the next 50 years, most scientists working on immunology focused on understanding the origin and character of the blood products that confer immunity. These blood components came to be called antibodies. The protective role of cells was ignored until the second half of the twentieth century.

We now turn to describing our modern understanding of the immune system, which will make clear that both antibodies and cells are critically important for our eternal battle against infectious viruses. Both blood and cells matter.

Current Understanding of Immunity

Our immune system is comprised of two inextricably linked parts called the innate and adaptive immune systems. The adaptive immune system generates a response that is tailored for the specific invading virus. This custom-designed response takes time to develop. If we relied only on adaptive immunity, by the time it was armed and ready for battle the invading virus would have multiplied many times and spread throughout the body, which would overwhelm us. The innate immune system prevents this from happening and keeps us alive until the adaptive immune system can kick in. It is an early warning system that alerts the body to a viral invasion. Sometime after infection you feel terrible because you have some combination of fever, body aches, inflammation, and loss of appetite. These symptoms are side effects of the actions that the innate immune system takes to fence in viruses near the site of infection and slow the growth and spread of virus particles. The adaptive immune system becomes fully active about a week after infection, and this is usually when you start feeling better. This is because your adaptive system has generated a response specifically tailored to the infecting virus, and, working together, innate and adaptive immunity are vanquishing the virus and eliminating it from your body. A memory of this victory is also established. Our story describing how all this works begins with how the immune system first encounters the infecting virus.

How the Infecting Virus Meets the Immune System

As we saw in chapter 3, to get into a cell, the proteins on a virus's surface bind to receptors on human cells. For example, the spike protein of SARS-CoV-2 binds to ACE2, its receptor on cells in the lung and other tissues. The virus then hijacks our own cell's machinery to assemble many new virus particles that are released from the host cell. These virus particles spread through the spaces in between the cells that make up tissues, looking for new cells to infect. A type of Metchnikoff's phagocytic cells, called a dendritic cell, lives in all tissues. Dendritic cells are like sentinels that constantly sample and ingest things in the environment in between the tissue cells. Most of the time, they ingest some waste product of a cell. But, if due to an infection, there are virus particles present, a dendritic cell will ingest them. When this happens, a dendritic cell pulls up its stakes and leaves the tissue. The fluid in the spaces in between cells in tissues is called lymph. Just as arteries and veins carry blood to and from our tissues, lymphatic vessels drain the lymph from tissues. Among other substances, the lymph fluid contains virus particles and the dendritic cells that are ready to leave the tissue because they have eaten viruses. At regular intervals, the system of lymphatic vessels is interspersed with organs called lymph nodes. They act as filters that retain viruses and dendritic cells in lymph nodes near the infected tissue. Cells of the adaptive immune system circulate in blood, and enter lymph nodes by squeezing through the walls of blood vessels that pass through them. This is how the adaptive immune system meets the virus in a lymph node.

Interactions between cells of the adaptive immune system, virus particles, and dendritic cells in a lymph node result in the generation of cells and their products that are like warriors equipped with specialized weapons designed to neutralize the effects of the specific infecting virus. These immune products migrate out of lymph nodes through lymphatic vessels that connect to the blood stream. They can then pass through blood vessel walls to enter tissues, where they prevent the virus from

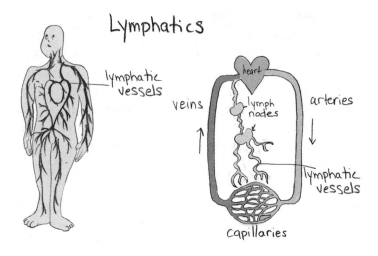

infecting new cells and kill the cells that are already infected. Antibodies are just one type of such specialized products that wage war with the virus. How antibodies specific for the invading virus are produced is the topic of the next section.

Antibodies: An Important Arm of Adaptive Immunity

After the discovery of antibodies to diphtheria toxin and other microbes, the obvious question was how humans could produce antibodies that were specifically tailored for neutralizing so many different disease-causing agents. Paul Ehrlich in Germany suggested that antibodies bind to their specific targets in the same way that a key fits a lock. The shapes of the antibody and the specific target that it binds to are complementary, just like the shapes of a key and a lock are matched. He then proposed that our cells' surfaces have all the antibodies necessary for specifically recognizing all the disease-causing microbes and toxins that commonly attack humans.

Lock and key model

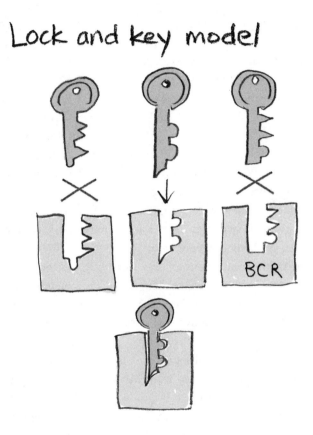

But this idea quickly ran into trouble because immunologists discovered that antibodies specific for chemical compounds were produced when these chemicals were injected into animals. As these compounds were not microbes or their toxins, this suggested that antibodies specific to almost anything could be generated. One idea proposed to explain this was that antibodies were flexible objects that could be actively molded into different shapes to bind to specific invading agents, including viruses and bacteria. But then how could memory for a previously encountered virus be retained? Why would the antibody maintain its shape after the virus was no longer present in the body?

In the 1950s, Niels Kaj Jerne in Denmark and Macfarlane Burnet in Australia proposed a possible solution to the puzzles noted above. In the short story "The Library of Babel," the Argentine writer Jorge Luis Borges describes a library in which there are books that contain every possible combination of letters. This library therefore contains every book that has been or could ever be written. Burnet proposed that the set of cells that comprised the immune system was like this library. Analogous to the sequence of letters in one book, each immune cell was equipped with an antibody that could bind to something different. So, we could mount immune responses specific to anything that we encountered in the past, present, or future. When a particular microbe or chemical invades us, the immune cell with the right antibody specificity binds to the invader. Binding causes this particular cell to multiply. The resulting progeny can then neutralize the infecting agent. Burnet also reasoned that memory of each past infection is imprinted in us because now the body would have many copies of the cell type with antibodies specific for the corresponding microbe. So, upon reinfection, the response would be rapid and robust because now, instead of only one cell specific to the microbe, many would be ready and waiting.

This model implied that every human is born with a huge diversity of immune cells, each with a different antibody. To be able to recognize almost anything specifically, each of us must have millions of different antibodies. But humans have only about 20,000 genes that encode information about all the proteins our cells can make. How could we have many more different types of antibodies than genes?

This puzzle was solved in 1976 by Susumu Tonegawa, a Japanese scientist then working in Basel, Switzerland. It had long been known that antibodies are produced by a type of cells called B lymphocytes, or B cells. B cells display a protein on their surface called the B cell receptor (BCR). Tonegawa discovered that the gene encoding information about the BCR is different in each B cell. This is because the BCR gene is comprised of different bits of DNA that have to be joined together to make a complete gene. Our DNA contains many flavors of each of these bits. One flavor of each bit is randomly picked and joined to assemble the BCR gene of a B cell. A different combination of bits is picked for each B cell, and so an enormous diversity of B cells with different BCRs can be generated. Indeed, each of us has about 100 billion B cells, and there are at least as many as 10 million types of B cells, each with a distinct BCR. So, most B cells have a BCR that is distinct from that of other B cells. Tonegawa's finding was very surprising because until then it was believed that every cell in a person's body had an identical copy of DNA. Tonegawa discovered that this was not so for B cells, and he was awarded a Nobel Prize in 1987 for this important discovery.

What does Tonegawa's discovery about B cells with different BCRs have to do with antibodies? When a B cell meets a virus particle in a lymph node, if its BCR can bind sufficiently strongly to a part of this virus's spike, then chemical reactions occur inside the B cell that cause it to start multiplying. The progeny secrete a soluble form of their BCR, and this is what we call an antibody. So, BCRs are basically antibodies displayed on the surface of B cells, just as Burnet and Jerne had proposed. It is remarkable that the insights of Burnet and Jerne based just on their imagination turned out to be largely correct. Burnet was awarded a Nobel Prize in 1960, and Jerne won his in 1984.

The secreted antibodies circulate through the blood and all tissues of the body, searching for the infecting virus that led to their production. An antibody binding to the spike protein on the virus can mask the spike, thus preventing the virus from attaching to receptors on human cells. Since this process neutralizes the virus's ability to infect healthy cells, such antibodies are called neutralizing antibodies. Antibody-bound viruses are destroyed, either because they are eaten by phagocytic cells or because chemicals in our blood can bind to the antibodies and punch holes in the virus surface.

A puzzling thing happens as the infection progresses that was first demonstrated by the late American physician and scientist Herman Eisen. Eisen spent part of World War II serving as a doctor on a navy ship. He did not have much to do sometimes, and so he read books on immunology. He found the topic fascinating, and after the war, he increasingly became interested in immunological research, and less so in practicing medicine. In experiments carried out with rabbits, Eisen and colleagues showed that the antibodies became more potent as time ensued after infection. Indeed, they found that antibodies could increase their potency over 1,000-fold in 1–2 weeks. Burnet's ideas could not explain this increase in antibody potency.

The answer ultimately came from Darwin's ideas on evolution. Darwin's landmark studies described how species continuously evolve to become fitter, or better suited to their environment. This process occurs because mistakes are made when our genetic material is copied, leading to mutations. Darwin's theory of evolution says that, over time, individuals with mutations that confer traits that enhance fitness take over the population. B cells undergo such a Darwinian evolutionary process during the first 1–3 weeks after infection. B cells that bind sufficiently strongly to a virus get activated and multiply in lymph nodes. During this process, mutations arise in the BCRs of the progeny of the activated B cells at an unusually high rate. The new B cells with mutated BCRs compete with each other to bind to the virus that caused the parent B cells to get activated. B cells that bind better survive, while the others die. Some of the B cells that survive leave the lymph node and secrete antibodies. However, the vast majority of the B cells that survive remain in the lymph node for further rounds of mutation and selection. The repeated rounds of mutation and selection result in B cells with BCRs that bind increasingly more avidly to the virus's spike. The corresponding antibodies that they secrete are thus more potent in masking the virus's spike and neutralizing its ability to infect human cells.

The antibody molecule has a Y shape with two identical binding sites for the virus spike. The stem of the Y-shaped antibody can be of different types, each of which can enable different antibody functions. During an infection, the first antibodies that are produced have stems of a type called IgM. As the infection progresses, the antibody response often changes to another type of antibody called IgG. There are other types of antibodies with different functions. For example, an antibody type called IgA specializes in protecting the surfaces of the body and is secreted into the gastrointestinal tract and the airway surfaces of the lung, mouth, and nose. Here they bind to and block viruses from attaching to and entering cells.

Antibody Tests and Their Significance

As described above, potent antibodies specific for the virus emerge 1–2 weeks after infection. Even after the infection is cleared from the body, these antibodies usually continue to circulate in the blood and tissues for a period of time (see more

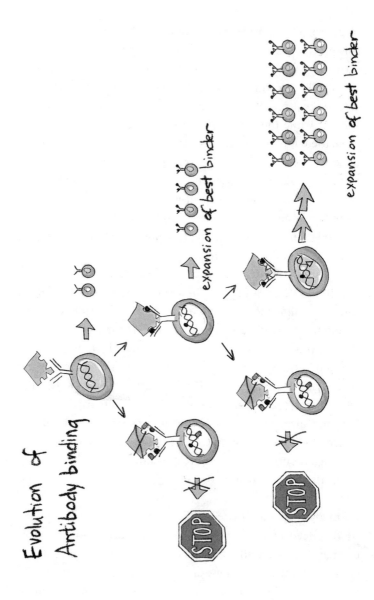

Evolution of Antibody binding

expansion of best binder

expansion of best binder

expansion of best binder

STOP

STOP

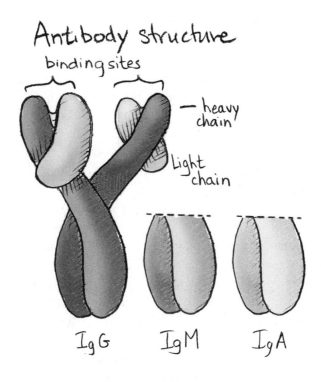

Antibody structure

binding sites

heavy
chain

Light
chain

I$_g$G   I$_g$M   I$_g$A

on the duration below). This is why a common way to screen individuals for a viral infection is to test whether they have antibodies specific to that virus in their blood. For example, the HIV screening test determines whether antibodies specific to the HIV virus are present. These so-called serological tests can be highly sensitive and specific, relatively simple, and inexpensive. The basic procedure is to mix a blood sample with the virus's proteins. If antibodies specific to the virus are present, they bind to the viral proteins. The bound antibodies can be detected by using a second antibody that binds to all human antibodies. There are many ways to perform this test. In one method, the virus's proteins are attached to a gold bead. If anti-viral antibodies are present in the blood, they bind to the bead. The mixture of blood and gold beads flows past a surface on which the second anti-human antibody is attached. If the anti-viral antibody is present, the gold beads are captured as the anti-human antibodies bind to the antibodies on them. The captured gold beads are easily visualized. Commercially made tests are available to test for antibodies specific for many microbes. Examples include measles, HIV, hepatitis C, tetanus, diphtheria, and SARS-CoV-2.

During the COVID-19 pandemic, it was critical to diagnose potentially infected people quickly to determine how widespread the disease was and to identify whether an acutely ill or hospitalized patient was infected with the virus. At early stages of infection, specific antibodies are not likely to be detectable. That is why during the COVID-19 pandemic, direct detection of the virus using the method described in chapter 3 was at first the preferred way to test for infection.

Later, as the pandemic progressed, antibody tests were deployed to identify those who were infected in the past but may not have been tested for the virus. As we will see in the next chapter, after the initial phase of a pandemic it is important for public health officials, employers, and the public to know how many have recovered from the disease and thus may be immune to reinfection. The accuracy required for a test to be useful for this purpose depends on circumstances. Antibody tests have to be specific. A test that produces false positives would mislead individuals, the public, and public health officials into believing that an individual or a large proportion of the population has recovered from the disease and is likely immune. The required accuracy of a test depends on the proportion of the population

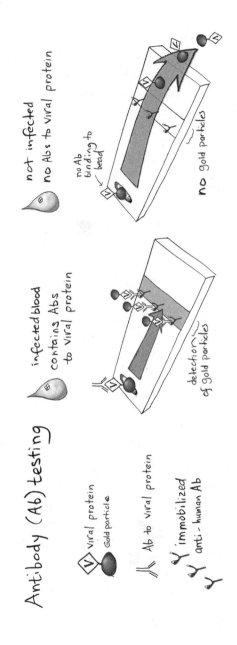

that is infected. Suppose that a particular vendor's antibody test has a 5 percent error rate. If 90 percent of the population is antibody positive, this test may be adequate for estimating the fraction of infected people and for guiding public health policy. However, if only 10 percent of the population is antibody positive, a more accurate test is likely required.

Recall that IgM antibodies arise early in the infection, followed by the more potent IgG antibodies. Thus, IgM type antibodies by themselves are usually an indicator of early infection. The presence of sufficient numbers of IgG antibodies usually signals that the person is likely to be protected from reinfection. COVID-19 is a disease of the respiratory tract, and so it may also be important to test for the presence of IgA antibodies because they protect surfaces in our lungs, mouth, and nose. Indeed, some of the antibody tests being developed now for SARS-CoV-2 infection status look for IgM, IgG, and IgA antibodies.

For reasons that are not properly understood, depending on the infecting virus, antibodies circulate in our blood and tissues for varying periods of time after recovery. For some viral infections, protection is conferred for the lifetime of an individual who has recovered. Available data suggest that antibodies generated upon infection with the virus that causes the disease SARS, which afflicted East Asia in 2003, circulate in recovered persons for up to two years after recovery from infection, but not much longer. While the SARS-CoV-2 virus that causes COVID-19 is closely related to the virus that causes SARS, it is not yet known for certain how long protective antibody levels circulate in persons who have recovered from COVID-19. For any viral infection, issues like this are clarified as more patient data become available.

Circulating Antibodies Are Not the Only Way to Retain Memory of Past Infections

The absence of circulating antibodies does not mean, however, that a person who has recovered from an infection cannot rapidly ramp up production of protective antibodies shortly after reinfection. After a person clears an infection, most of the B cells that were produced by the rapid multiplication of B cells in response to the particular virus die. This feature ensures that your immune system is not strongly biased to combat the virus that you just vanquished, and this is important because the next infectious microbe that assaults you is likely to be different. Also, given that we constantly battle infections, if the B cells that grow to large numbers during every infection did not die, we would soon become one large B cell! However, some of the B cells that were produced in response to the infection that was just cleared remain as so-called memory cells, which can rapidly produce antibodies and mount a robust response upon reinfection with the same virus. Memory B cells are not detected by antibody tests. So, the persistence of circulating antibodies specific to the virus in the blood is not the only way that memory of past infections is imprinted in our immune system.

T Cells: The Other Important Arm of Adaptive Immunity

As we described, antibodies principally attack free virus particles in blood or in the spaces in between cells in tissues. But this means that infected cells and the virus particles that they harbor are protected from antibody attack. To clear the infection, we also need to destroy the infected cells. T lymphocytes, or T cells, kill infected cells. Because of the intense focus on antibodies in the first half of the twentieth century, how T cells function was understood only more recently.

An organ transplanted from one animal to another of the same species is usually destroyed in the host animal, and this was a barrier for organ transplantation in humans. In the 1930s and 1940s, George Snell, an immunologist working in Bar Harbor, Maine, on the isolated and beautiful Mt. Desert Island, was trying to understand whether transplant rejection had a genetic origin. His strategy was to transplant organs from one purebred mouse to another. While organs transplanted between mice of the same strain were not rejected, organs from a different mouse strain were. He then interbred these strains of mice to identify the genes responsible for rejection. These experiments required breeding many generations of mice and took many years to complete. Perhaps the isolation of Bar Harbor and the long winters allowed Snell to have the patience to carry out these experiments. He eventually found that transplanted organs were rejected because of differences in a single group of genes. In humans, these genes are called HLA genes, and they must be compatible to prevent transplant rejection. This is why it is standard procedure now to match the HLA types of organ transplant recipients and donors. It took a while, however, to figure out why differences in these genes mattered.

The first clue as to the function of these genes was provided by the Australian immunologist Jacques Miller. The thymus is an organ located near the heart, and it was thought to be unimportant because removing the thymus of an adult mouse had no effect. At that time, the only known purpose of the thymus was in the context of sweetbreads, a delicacy in European cuisine that is prepared using the thymus of an animal, usually lamb. Miller found that he could prevent mice from rejecting transplanted skin by removing their thymus at birth. He also found that these mice were more prone to infections. Thus, he deduced that an important immune function must require the thymus, especially early in life.

We now know that T cells undergo a number of maturation processes in the thymus before they start circulating through our body. The "T" in T cells refers to the thymus being the site where mature T cells are produced. Miller found that T cells kill the cells of an organ transplanted from a mouse with different HLA genes, and this destroys the organ. Mice with their thymus removed at birth were missing T cells, and so transplants were not rejected. Miller's observation that these mice were also more prone to infections implied that T cells were an important part of our army of immune cells. Interestingly, shortly after puberty, the thymus begins to wither away, and thereafter we live largely with the T cells we have. By the way, this explains why all chefs are taught that sweetbreads are best prepared using the thymus from a young animal and that the adult animal's thymus is small, fatty, and inedible.

By the 1970s it was understood that T cells also kill virus-infected cells, but not free virus particles. But how they identify infected cells and thus know what to kill, and whether HLA genes had anything to do with this, was unknown. In a relatively short period of time various studies came together to provide us with our current understanding of how this works.

Hugh McDevitt and Michael Sela, working at the National Institute of Medical Research in England in the 1960s, were studying immune responses in mice injected with a specific protein. Using some of the mouse strains developed by Snell, they found that the ability to mount a strong or weak immune response to a specific infecting substance depended on a mouse's HLA genes.

Expanding on the work of McDevitt and Sela, Rolf Zinkernagel and Peter Doherty working in Australia discovered another link between HLA genes and the T cell response. They were studying the ability of T cells to kill cells infected with a particular virus. They collected T cells that were able to kill virus-infected cells from a mouse. They then tested the ability of these T cells to kill virus-infected cells from mice with different HLA genes. They found that T cells could kill infected cells only when the infected cell came from a mouse that shared the same HLA genes. That is, mice could kill their own infected cells but not those from mice with different HLA genes. This was a huge surprise. The result suggested that to recognize a foreign invader (the virus), the T cells needed to first detect that its own cell, and not that of another with different HLA genes, was infected. But why was this the case?

It was reasonable to assume that something on the surface of T cells was responsible for detecting infected cells, much as antibody-like B-cell receptors on the surface of B cells bind to the spike proteins of a virus. Soon, James Allison, John Kappler, Philippa Marrack, and Ellis Reinherz presented data from three independent studies that showed that indeed a protein on the surface of T cells was important for detecting infected cells. It was assumed that identification of this T-cell receptor (TCR) would explain why T cells from one mouse could kill virus-infected cells from another mouse only if they shared HLA genes. Because of this scientific importance, it was thought that the discoverer of this protein, and the gene that encoded it, would win a Nobel Prize. A mad race to find it began.

The race was ultimately won by two groups of scientists who used a novel approach: Mark Davis and Steve Hedrick, working together at the National Institutes of Health, and Tak Mak, at the University of Toronto. Their experiments were cleverly designed to identify the genes that only a T cell possesses. One of these genes would correspond to the T cell receptor. These experiments had to be conducted with precision—Davis has likened them to the precision he had to exhibit as a fencer in college. On a Sunday in 1983, Hedrick was on the way to the zoo with his family and stopped in the lab to check on the experiments. The results he saw made him realize that they had discovered the identity of the TCR. He proceeded to the zoo and a bit later called Davis to tell him that something special had happened. They did not know that, at the same time, Mak had similar results. The two papers describing the TCR were published together in 1984. These results also showed that the TCR gene was different in each T cell, and the reason was that, just as we described for the BCR gene, it was made up of bits of DNA that were randomly shuffled and joined. So, along with our supply of B cells, we also have an equally diverse repertoire of T cells as well.

Peering at the TCR protein and gene, immunologists were hoping to find the answer to the Zinkernagel–Doherty puzzle. But because the T cell receptor looks very similar to the antibody molecule, no new insights emerged regarding why T cells need to recognize self to recognize foreign. No one has been awarded a Nobel Prize for the discovery of the TCR.

Other studies provided the answer. While working with Brigitte Askonas at the National Institute of Medical Research in England, Emil Unanue, a Cuban émigré to the United States, made a surprising observation. Adding a drug, chloroquine, to target cells prevented T cells from detecting them. Unanue was subsequently offered a faculty position at Harvard, where he followed up on this observation.

Our cells have a mechanism to destroy used proteins. Used proteins are first chopped into fragments called peptides. These peptides are then further degraded into amino acids. Unanue's experiments showed that the peptides created during the processing of used proteins bind to the proteins encoded by the HLA genes. He then found that these HLA-bound peptides are sent to the cell surface, and they are the molecular flags of infection that T cells recognize. It turns out that one of many things chloroquine does is block the ability of cells to break down used proteins. So, cells treated with this drug could not be detected by T cells. When Unanue presented his results for the first time in 1984 at a scientific conference, he was greeted with derision, with one immunologist comparing the ideas to cells displaying their own feces on their surface.

Shortly after, in the mid-1980s, Pamela Bjorkman, a graduate student working toward her PhD in the laboratory of Donald Wiley at Harvard University, provided further evidence supporting Unanue's findings. Wiley's scientific specialty was structural biology. He studied the functions of proteins by coaxing them to form crystals that can then be imaged using X-rays to reveal what they look like. Bjorkman decided that her PhD research would focus on crystallizing and imaging one of the HLA proteins. After challenging efforts, she finally succeeded. The structure of the HLA molecule that Bjorkman and Wiley reported shows that it has a groove, and that in the groove lies a peptide, exactly as predicted by Unanue.

These findings, along with a few other facts, provided a solution to the Zinkernagel–Doherty paradox. Each one of us has between 6 and 12 different types of HLA proteins that are displayed on the surface of almost all our cells. These proteins are the most variable in the human population. Only identical twins and some siblings have exactly the same set of HLA proteins. Each HLA protein binds distinct subsets of peptides. To see why it is important to have so many different variants of

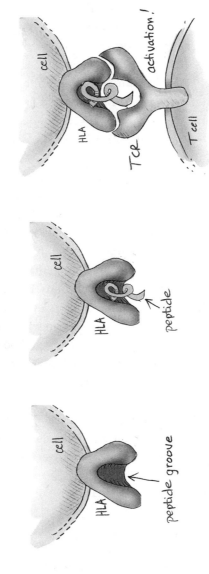

T cells see viral peptides bound to HLA proteins

HLA proteins in the human population, consider the following extreme scenario. If we all had exactly the same HLA protein, and one virus emerged whose protein fragments did not bind to its groove, none of us would be able mount a T-cell response to this microbe. Thus, this virus would pose a threat to the entire human species because none of us would be able to kill cells infected by it. Having many variant HLA proteins in the human population likely evolved as a bet-hedging strategy that prevents such an eventuality.

One more fact needs to be added to close the loop on understanding how T cells work. The thymus is like a school for baby T cells. Only some of the baby T cells that enter the thymus pass the requisite tests to graduate. T cells are trained to detect their own HLA proteins, and not bind to their own HLA-bound peptides too strongly. Thus, the T cells that exit the thymus have TCRs that bind with modest affinity to peptides derived from one's own proteins bound to one's own HLA proteins. T cells that bind too strongly or weakly are eliminated. When a mature T cell binds strongly to an HLA-bound peptide, it infers that this peptide is of foreign origin and not derived from the organism's own proteins. So, the cell displaying this HLA-bound peptide must be infected and should be killed. In Zinkernagel and Doherty's experiments, T cells in one mouse could not kill virus-infected cells derived from another mouse because these T cells were not trained to bind to and detect the HLA proteins of the second mouse.

T cells play a critical role in directing the immune response against viruses. When the TCR on a T cell binds strongly to an HLA-bound viral peptide displayed on a phagocytic cell that has eaten the virus, the T cell gets activated. Activated T cells begin to multiply, and the progeny go into tissues seeking infected cells. In tissues, if a T cell encounters a cell displaying the same HLA-bound viral protein fragment that originally activated it, the T cell can kill the infected cell. After the infection is cleared, some of these T cells live on as memory T cells, which can be quickly reactivated upon reinfection. It turns out that in addition to these killer T cells, there is another subtype of T cells. These T cells have many functions important for combating viral infections. For example, they secrete chemicals called cytokines, which have functions we will discuss in the next section.

Innate Immunity

As we recounted, the twentieth century was an age of great discoveries about our adaptive immune system. These discoveries taught us how our immune system does amazing things like mount virus-specific responses and establish memory of past infections that allows it to usually swat away viruses that reinfect us. The phagocytic cells that Metchnikoff had discovered were also found to be important for eating up all types of viruses present in tissues, transporting them to lymph nodes, and displaying HLA-bound viral protein fragments that activate T cells. But as the twentieth century progressed, it became clear that B cells, antibodies, T cells, and phagocytic cells could not be the whole story.

It takes about 5–10 days for the adaptive immune system to identify the rare B cells and T cells specific for a particular virus and mobilize them for active duty. But early in the infection, the virus multiplies and infects new cells. By the time the adaptive immune system is mobilized to engage the enemy, the virus will likely have already won the battle. As we usually win battles with viruses, something else must be happening during the early stages of infection when time is of the essence. Controlling the infection early requires another part of the immune system that is ready to go right away upon infection, and as we mentioned early on in this chapter, this part of the immune system is called the innate immune system.

As immunologists started to consider how the immune system reacts immediately after infection with any microbe, Charles Janeway, at Yale, started talking about the immunologists' "dirty little secret." What immunologists did not talk about was that if you injected a foreign protein or chemical into an animal, nothing happened. To induce an immune response, they had to mix in mysterious substances like dead bacteria, oils, or some other noxious material. For decades it was also known that for a vaccine to elicit an adaptive immune response that protects a person from a particular infection, something similar had to be added to get the immune system going. Janeway and others reasoned that there must be more to innate immunity than just phagocytosis. They proposed that cells of the innate immune system must have receptors that could recognize some features present in all bacteria, fungi, and viruses. Detection of these features would tell the immune system to mobilize.

Evidence for such a system of cells came from studies in insects in the 1990s. A French biologist, Jules Hoffmann, was a specialist in studying grasshoppers and fruit flies. These organisms do not have T cells and B cells, and he wondered why they did not die of infections. He repeated Metchnikoff's experiments and began to think about how cells of the innate immune system are alerted to go to the site of infection. To try and understand this, he mutated flies and looked for ones that died of overwhelming infections because of faulty immune systems. His studies showed that a receptor identified earlier by Nobel Prize–winning scientist Christiane Nüsslein-Volhard, in a different context, was also important for preventing infections. Nüsslein-Volhard studied genes that were important for proper formation of the fly body. Flies with mutations in one of these genes had bizarre and unusual shapes. Seeing these unusual flies for the first time under a microscope, she is said to have exclaimed "Toll!" ("Cool!" in German) and named the gene "Toll." The idea that a gene implicated in controlling body shape would also function to mediate immunity in flies seemed hard to believe, but Hoffmann and his group proved that Toll did just that.

Around the same time, Bruce Beutler, working at the University of Texas Southwestern Medical School in Dallas, was studying severe bacterial infections. As a graduate student in New York, he helped to identify a substance called tumor necrosis factor (TNF) that is produced upon overwhelming bacterial infection, resulting in a condition called septic shock. For decades, it was known that a component of bacteria called lipopolysaccharide (LPS) was a powerful stimulator of TNF. When he started his own laboratory in Dallas, Beutler decided to search for the receptor for LPS and discovered that it was a human version of the Toll protein Hoffmann was studying. Beutler and Hoffmann would win Nobel Prizes for their work in 2011.

Ruslan Medzhitov, at Yale, and others have identified a large family of Toll-like receptors, and how they function, in mice and humans. Each of these receptors binds to different components that are specific to bacteria, fungi, and viruses. These discoveries showed that these receptors not only recognize microbial pathogens but also tell the body which type of microbe has invaded. The receptors that respond directly to bacteria, fungi, and viruses goes far beyond the Toll receptor family, and currently over 50 such receptors have been identified. This innate immune system is present in many species in nature, and functioned to protect living organisms long before vertebrates (like us) acquired an adaptive immune system.

When innate immune cell receptors bind to a component of a microbe, the cells respond by producing hormones called cytokines (which T cells also produce). Cytokines signal to the body that a dangerous microbe is present and create an environment that is inhospitable for the invader. The effects of cytokines include increasing body temperature, inducing the liver to produce massive amounts of antimicrobial proteins, changing our metabolism so we can survive without much nutrition, and mobilizing immune cells. Fever helps because most microbes do not replicate well at higher temperatures. The antimicrobial proteins from the liver bind to and damage the infectious agent, and the metabolic changes cytokines induce (that make sick people lose appetite) aim to starve microbes. There are also cytokines called interferons that tell cells to turn on genes that specialize in blocking viral replication. It is the cytokines produced by innate immune cells that make you feel sick when a virus infects you.

The activated innate immune system is a blunt weapon that slows virus growth and spread, which allows us to survive until the adaptive immune system can take over. In some cases, the innate immune system is able to eradicate the invading virus by itself. The activation of innate immune cells is also required to release the functions of the adaptive immune system. This is why immunologists had to add noxious chemicals and bacteria to mimic a real infection when they wanted to elicit a T-cell or B-cell response.

Over a hundred years ago, the humoralists won the day. Today, we know that Metchnikoff's phagocytes are central to the immune system. The importance of the cells of the innate immune system for protecting us from infectious diseases is made vivid by the fact that humans who lack an important phagocyte, neutrophils, do not survive beyond a few weeks without major therapeutic interventions. Metchnikoff's Nobel Prize in 1908 turns out to be prescient and highly deserved.

But innate immunity can also go awry. The reason why some patients have bad outcomes, including death, when they are suffering from COVID-19 is because of misregulation of innate immunity. This leads to something called a cytokine storm, which is characterized by uncontrolled secretion of cytokines that just does not turn off. SARS-CoV-2 infects organs in the respiratory tract, like the lungs. An overly exuberant immune reaction is especially perilous in the lung, where oxygen transfer into the blood occurs across thin membranes that can be damaged by cytokines. Some reports even suggest that SARS-CoV-2 does something very insidious. It turns off the production of interferons, the cytokines that block viral replication, while promoting the production of cytokines that cause other effects of inflammation, leading to a cytokine storm. As we learn more, we hope we will develop good therapeutics to help with such conditions. Common autoimmune diseases, such as rheumatoid arthritis, lupus, or inflammatory bowel disease, are also exacerbated by an overexuberant innate immune system. Drugs that try to block these effects have been developed, and indeed cytokine blockers are among the best-selling drugs on the market.

Putting It All Together

Our first line of defense against infectious microbes is our innate immune system. Cells of the innate immune system destroy microbes by ingesting them and by secreting chemicals such as cytokines. The innate immune system is remarkably efficient, and many viral invasions are eliminated by innate immunity. When patients are infected with SARS-CoV-2, which causes COVID-19, there are no symptoms for a few days. Up to 50 percent of those infected are either asymptomatic or have mild symptoms. Most likely, for the first few days after infection, and in those with mild or no symptoms, the innate immune system is able to largely control the virus through secretion of cytokines. For example, interferons can suppress viral replication.

The onset of symptoms is a sign that the innate immune system is fully activated, and this begins mobilization of the adaptive immune system. Many different cytokines are produced at this stage and patients feel very ill. Because SARS-CoV-2 infects the lungs, there is inflammation in the lung and this impairs the ability to transfer oxygen from the air to the blood, leading to respiratory distress. Around 5 to 7 days after the beginning of symptoms, patients appear to either get better or get much worse. The timing is consistent with the adaptive immune system kicking in.

B cells are a key component of adaptive immunity. Each B cell displays a receptor, its BCR, on its surface. The BCR on one B cell is likely to be distinct from the BCR on another B cell. If the BCR on a particular B cell can bind sufficiently strongly to a part of the spike proteins of a particular virus, then it begins to multiply. A Darwinian evolutionary process then takes place, which ultimately leads to the secretion of soluble forms of the BCR that bind even more avidly to the virus. These are antibodies. Antibodies migrate to tissues and bind to the infecting virus, preventing them from infecting cells. Phagocytic cells of the innate immune system also ingest and digest the antibody-bound viruses. There are many kinds of antibodies that arise during different stages of infection, and they combat the virus in different ways. These processes lead to IgM and then IgG antibodies becoming detectable in blood about 5–7 days after the beginning of symptoms.

But antibodies principally combat virus particles in blood or in the spaces between cells in tissues. T cells wage war against infected cells. Most T cells have a distinct receptor (TCR). Infected cells display protein fragments derived from the virus bound to our HLA proteins. If the TCR on a particular T cell can bind sufficiently strongly to the HLA-bound viral protein fragment on an infected cell, it gets activated and multiplies. The activated progeny migrate to tissues. In tissues, one class of activated T cells, killer T cells, kill infected cells that display the same viral protein fragment that originally activated the parent T cell. They do so by secreting chemicals that punch holes in the infected cell. Because the SARS-CoV-2 virus infects cells that have the ACE2 receptor, successful clearance of the virus likely involves killer T cells. In addition to killer T cells, there are other subtypes of activated T cells that perform functions like secreting cytokines and assisting B cells in the production of high-affinity antibodies.

Why do some patients infected with SARS-CoV-2 do more poorly than others? It is possible that in some patients, the immune response is defective or is activated too late to control the widespread growth and spread of the virus. Sometimes when the immune response senses incorrectly that the infection is out of control, it mounts an immune response that is out of proportion to the threat. Cytokines send out a general alert of infection. But when the immune system senses the potential that the body is in grave danger, immune cells can start to produce levels of cytokines that are too high, leading to a cytokine storm.

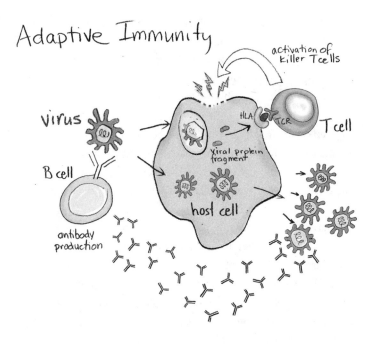

This is especially precarious for respiratory infections, as inflammation due to cytokines causes swelling and fluids to leak into the air spaces in the lung, making it difficult to breathe. A cytokine storm can also cause the blood pressure to drop, putting the body into shock. In such a condition, uncontrolled blood clotting occurs and multiple organs begin to fail. While it may be possible to treat this condition by suppressing the immune system, in practice, it is difficult to do so with precision. This is because it is difficult to distinguish between the parts of the immune system that are fighting the virus from the parts that are threatening the survival of the patient.

If and when a person successfully clears an infection, most of the T and B cells that multiplied in response to the causative virus die. But a few remain as memory T and B cells, and antibody levels are elevated for some time. These ready and waiting warriors allow a rapid and robust response upon reinfection. This virus-specific immunological memory is the basis for vaccination. A vaccine aims to elicit memory T and B cells, and antibodies that are specific for the virus against which one wishes to protect the population. We will say much more about vaccines in chapter 7.

5   Spread and Mitigation of Pandemics

In Boccaccio's famous novel, *The Decameron*, seven women and three men flee to a villa outside Florence to wait out the bubonic plague epidemic in 1348. To pass their time, they devise a system of storytelling, each being assigned to tell a story five nights a week for 2 weeks. Boccaccio ends the novel without telling us whether 2 weeks was enough time to let them return to their homes and resume normal lives. As weeks of lockdown continued during the COVID-19 pandemic, many of us felt as if we had just finished the *Decameron*, as we were left wondering when we could return safely to our normal lives.

How do we know how quickly a viral infection will spread? What steps do we need to take to mitigate the spread? When do we know that the peril has either passed or can be managed? These questions are very difficult to answer. In this chapter, we will describe the most important factors that influence the answers to these questions. We will also describe the essence of mathematical models that are used by epidemiologists to help guide public policy decisions during pandemics such as the COVID-19 crisis. These models can be useful for qualitative comparisons of the effect of different epidemic mitigation strategies on future outcomes. It is difficult, however, for these models to make numerically precise projections, especially in the early stages of an ongoing pandemic when the data are sparse and noisy.

The most important factors that influence whether a virus will cause a frightening infectious disease pandemic are how infectious the virus is and how fatal is the disease it causes. Let's explore these concepts.

Mortality Rate

In February 2003, a Chinese medical professor from Guangdong Province traveled to Hong Kong. After checking into his hotel, he began to feel ill. He was admitted to a hospital, where he died two weeks later. Unknowingly, he infected up to 23 other individuals in the hotel, who later traveled to Singapore, Toronto, and Hanoi and initiated new rounds of infections. The person who traveled to Hanoi was an American businessman who was urgently hospitalized soon after arrival. Concerned that he might be infected with a new virus, his doctors in Vietnam contacted the World Health Organization (WHO). The WHO sent an Italian physician, Carlo Urbani, to examine the patient. Urbani, who received the Nobel Peace Prize when he worked for Doctors without Borders, was alarmed by what he saw, and alerted the WHO of the possibility that a new virus might be circulating. Unfortunately, Urbani contracted the disease and died in Bangkok on March 29. In April 2003, the US CDC and Canadian authorities announced that they had isolated and identified a new virus. Thus began the SARS pandemic, which is caused by SARS-CoV, a coronavirus closely related to SARS-CoV-2. This virus was frighteningly deadly. SARS had a mortality rate of 10 percent—that is, 10 percent of infected people died. Another related coronavirus, MERS, which emerged in 2012, has a mortality rate of about 35 percent. One of the most lethal viruses is Ebola, which circulates in Central and Western Africa and has a mortality rate of more than 50 percent.

Viruses that cause diseases with a high mortality rate place a tremendous burden on the healthcare system. This is because almost everyone these viruses infect becomes seriously ill and requires hospitalization, often in the intensive care unit. The demand for care can overwhelm the capacity of the healthcare system, which, combined with the high mortality rate, results in a frightening situation. This problem is exacerbated by healthcare workers being especially at risk. This is because the virus's lethality is not known in the beginning, and the need for special protective equipment for the safety of healthcare workers is not yet recognized. Early warning about the mortality rate of a viral infection is critical for protecting healthcare workers and taking steps to mitigate the spread of the disease.

Even though highly lethal viruses are frightening, they usually do not spread widely and cause global pandemics. This is because infected people quickly become too sick to move around and infect others. The spread of the virus is thus mostly limited to healthcare workers and family members. So, the outbreak is usually easily contained. At the other end of the spectrum are viruses that cause only a mild illness and almost no deaths. Because infected individuals do not feel terribly sick, they go about their daily lives and can spread the virus to many with whom they interact. But since almost no one falls very ill, these viruses are just nuisances, and not a public health hazard. Four common coronaviruses that cause about 30 percent of common colds are examples.

SARS-CoV-2 is a virus with multiple features that are ideal for causing a global pandemic. It has a low to moderate average mortality rate, estimated to be roughly 0.3–1 percent. This is much lower than the mortality rate for SARS, MERS, and Ebola, but higher than for seasonal flu (roughly 0.1 percent). Most infected people have mild to moderate symptoms or none at all, and the onset of symptoms of illness occurs several days after being exposed to the virus. This makes SARS-CoV-2 tricky because infected people can transmit the virus to others before exhibiting any symptoms. Since many infected people do not feel very sick when they are transmitting the virus, they go about their normal lives and spread the infection to others. With many people infected, even if a fraction of patients require hospitalization, the capacity of healthcare systems can be overwhelmed, even in well-resourced countries. In the period between 2010 and 2019, the number of people who died of influenza in the United States ranged from 12,000 to 61,000 people each year. Since the mortality rate for the COVID-19 is perhaps ten times higher than that for influenza, if a comparable number of people are infected, the relatively moderate mortality rate for COVID-19 can still cause a frighteningly large number of deaths. To know how many people will be infected by a virus, we need to know how infectious it is.

Infectiousness of a Virus

The Concept of R_0

How quickly a viral infection can spread in a population is measured by a quantity called the basic reproductive number, which is abbreviated as R_0 (pronounced "R naught"). After a person is infected with a virus, there is a period of time over which they can transmit the infection to others. R_0 is equal to the average number of people an infected person infects during the infectious period when the entire population is susceptible to the virus. Suppose the infectious period of a virus is 10 days. To measure R_0, we could follow 100 infected individuals around for 10 days after each got infected, determine who they were in contact with, and find out how many of them became infected. For each person, the number of people they infected would be somewhat different because of differences in social networks (how many people they interact with), differences in types of interactions (e.g., outdoors versus inside a restaurant), random chance, and many other factors. We can, however, take the total number of people infected by the people we followed and divide it by 100 to get the average value of R_0. Why do we care about the value of R_0?

Suppose a particular viral infection is associated with a value of R_0 that is greater than 1—say, 2. So, one person infects two people, who then infect two others, and so on. How quickly the number of cases can grow in such a circumstance is illustrated by the likely apocryphal story of an Indian king who loved to play chess. He often challenged wise people and traveling sages to play with him, and offered them financial incentives. One day, such a person, who was an expert chess player, accepted the challenge. He told the king that, if he won, he would like only some grains of rice. The amount of rice would be determined by the king placing one grain of rice on the first square of the chessboard, two on the second square, four on the third square and so on—just like the growing number of infected people in our example. After the king lost, he ordered a bag of rice and started putting rice grains on the chessboard. Very soon, he realized that he had been duped. By the time he got to square number 64, the chessboard would be covered by more than 18,000,000,000,000,000,000 grains of rice! This type of growth is called exponential growth. So, if R_0 is greater than 1, a viral infection can expand explosively and can quickly overwhelm the capacity of the healthcare system and cause a lot of deaths. Many common childhood diseases are highly infectious. Measles has an estimated R_0 between 12-18, mumps has an estimated R_0 between 10 and 12, and chicken pox has an estimated R_0 between 10 and 12. In contrast, the 1918 pandemic-causing influenza virus had an estimated R_0 of about 2, smallpox an R_0 of roughly 5, polio between 5 and 7, and seasonal influenza about 1.5.

"Rice and the Chessboard" Exponential Growth

When $R_0 = 1$, the virus does not spread rapidly across the population. One infected person can spread the virus to only one other person during the infectious period. So, once the first person recovers, the total number of infectious individuals does not change. This stable situation is referred to as an endemic infection. When $R_0 = 1$, while the number of *infectious* people does not increase, the number of people who have been *infected* grows with time.

When R_0 is less than 1, one individual can, on average, pass the virus to less than one other individual. Since partial individuals don't exist, this means that one person may or may not pass the virus to another individual. For example, if we have a situation where the R_0 is one half, there are 50 percent fewer infected people per infectious cycle. So, when R_0 is less than 1, the number of infected people declines over time and eventually goes to zero, and the virus dies out.

The length of the infectious period is another factor that is important in estimating how quickly a virus spreads. If, for example, R_0 is 2 and the infectious period is 100 days, an infected person transmits the virus to one other person every 50 days. For the same R_0, if the infectious period is 10 days, a new person would be infected every 5 days. A virus characterized by a high value of R_0 and a short infectious period spreads frighteningly fast.

R_0 Is Not an Absolute Number

While R_0 is often discussed as if it is an absolute number for a given virus, it is not. Its value depends on many factors. A virus that is spread by human-to-human contact will have a higher R_0 in a city where people live close together than in a rural area where much less human contact occurs. This is why respiratory infections spread more rapidly in cities like New York, Paris, and London compared with rural areas. Social customs in different countries can also influence the value of R_0. In Italy, where multiple generations of families live together, R_0 will likely be higher in comparison to countries where nuclear families are the norm. R_0 can vary with the time of the year. For example, it could be higher in the winter because the virus is more stable in cool dry weather, and/or because people are confined indoors in the winter leading to more social contact in confined environments. This is true for influenza and the viruses that cause the common cold.

How quickly a virus spreads also depends on the stage of a disease epidemic. In early stages, most people have not been infected and so are susceptible to the disease. This is the stage when the number of infected people grows exponentially. As the infection progresses, an increasing number of people become infected, and some recover and are usually immune. So, the number of people susceptible to infection keeps declining. If an infected person typically meets 100 people during the infectious period, in the early stages of an epidemic all are susceptible. If R_0 for a particular virus in this social setting is 4, this person would infect four people. Now suppose that in the late stages of an infection 97 percent of people have recovered from the infection. Then, an infected person who encounters 100 people in the infectious period would interact with only three susceptible people, and would certainly not infect four new people. So, the virus would spread more slowly. While this example is contrived to make a point, you can see how the growth in the number of new cases changes with the stage of an epidemic by just looking at data for the COVID-19 pandemic. You can find this data on the website of the Institute for Health Metrics and Evaluation or the similar site maintained by Johns Hopkins University. The growth in the number of cases is exponential only for the first few hundred cases in any US county or any nation. After that the growth rate is slower because a newly infected person encounters fewer susceptible people. For example, if one person starts to spread the infection to their social network and another person does the same, soon their social networks overlap. The two people cannot both infect a common person in their social network.

So, it is fair to say that R_0 has an intrinsic value at the beginning of an epidemic when everyone is susceptible, and which depends on the virus itself, local living conditions, and social networks. But even in the same county or country, the rate at which the virus spreads changes with time. Therefore we should think about an effective R_0, or R_{eff}, that depends on time and various local conditions. We will discuss how R_{eff} depends on these variables soon, but can the intrinsic value of R_0, even in a particular county or nation, be measured accurately?

It is not easy to measure R_0 accurately for an epidemic caused by a new virus, as was made vivid in the case of the COVID-19 pandemic. When COVID-19 first began to spread in China, medical authorities did not initially realize that a new virus had emerged. It took some time to isolate the new virus and then to develop a specific test. By this time, basic information about the early stages of the infection was not available. Furthermore, imprecision in estimates of R_0 was likely exacerbated because the onset of the epidemic in China coincided with the Lunar New Year, a major holiday there. This accelerated travel (and viral spread) throughout China and encouraged close contact of those infected with their families.

When early efforts to control an epidemic fail, the biggest concerns are that the capacity of the healthcare delivery system will be overwhelmed by the spreading infections and that many people may die. The value of R_0 (or R_{eff}) is an important parameter that influences how quickly infections spread, and thus how quickly the number of hospitalized patients and fatalities will grow. In the absence of reliable data, can we estimate the value of R_0, and with some assumptions, make useful projections about the future?

Epidemiological Models

Epidemiologists use mathematical models to study how outcomes depend on various scenarios of virus infectivity, nature of local conditions, and policies imposed to mitigate the spread of disease. It is important to understand the essence of these models in order to know what they can and cannot do.

One of the simplest epidemiological models divides the population into four classes of people: (1) those who are susceptible (S) to infection because they are not immune to infection, (2) those who have been exposed (E) to the virus by coming in contact with an infected person, (3) those who progress to infected (I) status after exposure, and (4) those who have recovered (R) from the infection. For all viral infections that we know of, people who recover from a disease are immune for at least some duration. So, it is fair to presume that usually people in the "R" pool are protected from reinfection for a while. These models are called SEIR models for obvious reasons. The essence of these models is to describe the processes shown schematically in the accompanying figure.

The first step describes the rate at which infected people encounter susceptible people and expose them to infection. The second step describes the rate at which exposed people actually become infected. In order to describe how the number of infected people will grow, we need to know the rates at which this sequence of steps occur. The rates depend on the duration of the infectious period and R_0. As described above, the latter quantity depends on many factors, such as the social network of people, whether the population under consideration is rural or urban, or whether there was a special event occurring like the Lunar New Year in China or Mardi Gras in New Orleans (super-spreading events). The last step in the process describes the rate at which infected people recover from disease, which depends on the specific viral infection and traits of individual patients.

The various rates required to describe the processes noted above are called the parameters of the model. By knowing or assuming values of the parameters and the number of susceptible, exposed, infected, and recovered people at a particular time point, a well-established mathematical technique called ordinary differential equations can be used to calculate how the number of people in each of these groups will change with time. This is how epidemiological models try to project what the future will look like. Importantly, the parameters required to carry out these calculations are not firmly known for a new virus. R_0 is not usually known, and even the number of infected persons at a particular time is not known firmly. This is because without a very rigorous testing infrastructure in place, the number of infected people is difficult to determine. With uncertain values of the parameters, it is difficult to make accurate projections.

Epidemiological models can be made more complicated by adding many additional features that make them more realistic. For example, for COVID-19 one could divide infected people into groups of asymptomatic and symptomatic patients; divide those infected into who requires hospitalization, who does not, and who dies; or stratify the population by age, co-morbidities, geographic location, or other factors. You could also say that the rate at which an asymptomatic infected person transmits the virus is different from a symptomatic one or that patients who are hospitalized take longer to recover. For unknown reasons, some individuals spread the infection to many more than most people. One could try to account for the effects of such super-spreaders, who are characterized by values of R_{eff} that are much higher than the average. Every new feature added to epidemiological models correspondingly requires a new parameter to be specified. So, more complicated models have even more unknown parameters, making projections for future outcomes even more challenging.

Some of the unknown parameters can be estimated by adjusting their values until predictions from the model fit known data. For example, a model could tell us how the daily increase in the number of infected people or those who died changes with time. The parameters can then be adjusted to fit what was really observed for changes in new cases and deaths. As just one example, R_0 could be estimated this way. During the early stages of an epidemic, if tests are not readily available, the real-world data are both sparse and noisy. So, the parameters that are estimated in this way are not very accurate. For example, when testing is not widely available, the number of new reported cases is not an accurate reflection of reality. So, one might use only the reported deaths to estimate model parameters as these data are usually more reliable. But the uncertainty in the estimated parameter values is greater if less information is known, just like your answer to a question is less certain if you are given one clue rather than two. So, estimating many parameters accurately by fitting model predictions to match just the known number of deaths is more challenging than if the numbers of both deaths and new infections were reliably known. The more complicated models require estimating more parameters, which makes the estimates even more statistically unreliable. So, the projections that the models make are not numerically accurate. As more data are collected, the parameters are estimated more accurately and the projections get more accurate.

In spite of these challenges, models can be useful for making qualitative projections and estimating the relative effects of different public health measures on mitigating the spread of disease. In particular, models can be used to make projections for how public health measures may affect future hospitalizations and deaths, and the most important factors to control in order to keep these numbers low. This is very useful information for officials and leaders who have to make decisions with limited data during a spreading epidemic.

Effects of Public Health Measures on Mitigating a Spreading Epidemic

Testing, Quarantine, and Contact Tracing

The traditional method of epidemic control is to isolate those who are infected, identify all individuals they contacted, and quarantine them during the infectious period. Contact tracing requires a dedicated team to interview and track down all contacts during the infectious period. For logistical reasons, this is easier to do when the number of infected individuals is low. Therefore, this is a very effective way of epidemic control in the early stages of an epidemic.

Countries that receive some advance warning of a possible pandemic coming to their shores and become very vigilant can be very effective at controlling epidemics early on by extensive testing, quarantine, and contact tracing. During the SARS epidemic in 2003, early and quick action by authorities isolated those who were infected and identified the people with whom they came in contact. In the Hanoi hospital where Carlo Urbani raised the alarm that a new virus was present, the infection quickly spread to 40 hospital workers. Recognizing that this was a dangerous and infectious virus, the hospital staff locked themselves in to prevent the virus from spreading to the surrounding community. Vietnam closed its borders immediately, and a short six weeks later, with no new infections, Vietnam declared victory over SARS. Similar stories and timetables played out in Hong Kong, Toronto, and Singapore.

The characteristics of the virus that caused SARS helped make it easier to contain. It is a lethal virus, with most patients getting very sick, making it easier to identify those who are infected even without extensive testing. The peak infectious period of SARS-CoV-1 occurs after symptoms begin. Since hospitalization follows quickly after symptoms begin, spread within the community is limited. Hospital workers and close family members are the most vulnerable, and they can be quickly quarantined. The control and eradication of SARS, a virus with an R_0 now estimated to be about 3, within a few months of detection, is a powerful example of the effectiveness of isolation, contact tracing, and quarantine. But SARS-CoV-2 has very different characteristics compared with SARS-CoV-1, with many infected persons being asymptomatic and peak infectiousness at or before the onset of symptoms. Can isolation and contact tracing be a useful strategy for a virus like that which caused COVID-19?

Vigilance and quick action in terms of testing, quarantining, and contact tracing allowed some countries to manage the spread of the COVID-19 epidemic well. South Korea, Hong Kong, Vietnam, Taiwan, and Singapore were watching the growing epidemic in China with increasing trepidation. Almost immediately after China's announcement of an infectious outbreak caused by a new virus, these countries began monitoring the temperature of all passengers arriving on flights from China, especially from Wuhan. South Korea was able to identify its first case of COVID-19 on January 20, 2020, when a passenger from Wuhan arrived at the airport in Korea with fever. This person was isolated, and the country immediately began to ramp up testing nationwide over the next 2 weeks. Hearing this news, Taiwan restricted its borders to China on January 23 and started ramping up testing. Thus, both countries embarked on a program of extensive testing relatively early in the epidemic and then isolating infected persons, which limited the spread of the virus. Quarantine and contact tracing measures reduces the value of R_{eff} quickly. Early on, South Korea reported a value of R_{eff} less than 2 and Taiwan a value less than 1. These values of R_{eff} also reflect that the population of these two countries cooperated with social distancing recommendations.

The situation in the United States was different. That country detected its first case of COVID-19 in Washington State on the same day as South Korea, on January 20, 2020. This patient arrived on a plane from Wuhan a few days earlier, and went to a local clinic on January 19 with cough and fever because he was aware of the epidemic in Wuhan and thought it prudent to seek medical help. He tested negative for influenza, and because he didn't seem seriously ill, he was sent home while awaiting results of his nasal swab test for SARS-CoV-2 performed by the CDC in Atlanta. After the test came back positive on January 20, he was hospitalized and put into isolation and released two weeks later after clearing the infection. On January 31, 2020, the United States banned incoming travel from China. But US citizens could still enter the country after this date.

Over the next month, about 15 cases were identified sporadically across the United States in addition to 44 infected Americans identified on a cruise ship. In late February, a patient with no history of travel to China or any known exposure to an infected patient tested positive for SARS-CoV-2. This signaled a new stage of the epidemic in the United States. As this person's infection could not be traced to international travel, they were likely infected by someone in the local community. Epidemiologists call this stage of spreading of an epidemic "community spread." The CDC initiated a sentinel testing program with the goal of testing patients with mild respiratory illnesses for COVID-19 in six different cities. During the second week of March, 5 percent of patients tested in Chicago were positive for SARS-CoV-2. But a test was still not widely available, as the virus undoubtedly continued to spread. In mid-March, the CDC finally allowed tests other than its own to be used. The lack of widespread testing and contact tracing during the first 8 weeks or more of the epidemic in the United States made it difficult to contain the spread of the virus. Testing, isolation, and contact tracing were no longer possible once the virus started spreading widely.

The situation that prevailed in the United States during the early stages of the COVID-19 epidemic is not unusual for a virus with the characteristics of SARS-CoV-2. Viruses that have high numbers of asymptomatic infections and that can spread before the onset of symptoms are very challenging for public health officials to control. Because asymptomatic patients don't realize that they are infected, they can be identified only by extensive random testing and tracing all contacts of infected individuals. A long infectious period before the onset of symptoms is also challenging for contact tracing because of the limitations of human memory. It is difficult for people who test positive to remember every person they came in contact with some days ago. Some countries like Israel and Taiwan are using cell phone tracking technology to monitor movements and contacts. This is a potentially powerful approach for contact tracing, but with obvious privacy concerns.

When extensive testing, quarantine, and contact tracing are no longer possible because the virus is spreading rapidly, what other public health measures can be put in place to mitigate the spread of an epidemic caused by a highly infectious virus? The biggest concern at such a time is that the healthcare system may be overwhelmed leading to many unnecessary deaths. For various scenarios regarding hospital capacity, R_{eff}, and other variables, mathematical models can be used to compare the effects of different public health measures. Let us first consider the effects of social distancing, a strategy that was employed by many countries during the COVID-19 pandemic.

Social Distancing Can "Flatten the Curve"

The idea of keeping your distance from a potential infected person is likely as old as humanity. In biblical times, leper colonies were created to isolate lepers from coming in contact with healthy members of the community. As we mentioned earlier, during the bubonic plague in the fourteenth century, without even knowing that microbes caused disease, people fled from Florence to the countryside to wait for the disease to pass. Modern social distancing policies have their origins in the regulations put in place in St. Louis during the 1918 influenza pandemic.

In the summer of 1918, Dr. Max Starkloff, the health commissioner of St. Louis, was monitoring the growing influenza epidemic in Boston and its spread across the continent. In early October, the first cases identified in St. Louis were in a family of seven. The next day, when 50 more cases were identified, Starkloff sprang into action. He urged the mayor of St. Louis to forbid gatherings of large numbers of people and to close movie theaters, churches, pool halls, and concert halls. He also closed the public schools. As the number of cases grew, Starkloff began to restrict business activities. By early November, even though the number of new cases was beginning to stabilize, he moved to close all nonessential businesses, in spite of the emotional objections of the business owners. He may have done this to prevent a massive public gathering for the celebrations planned for November 11 to commemorate the end of World War I. An experiment to ease restrictions in mid-November resulted in a new outbreak of cases in children, and he reimposed strict closures of schools and businesses. As cases declined at the end of December, he gradually eased restrictions and allowed normal life to resume.

The concept of using social distancing as a public health measure was strongly supported by the St. Louis experience, which resulted in a lower fatality rate there compared with other cities during the months of September 1918 to February 1919. An oft-made comparison is with what transpired in Philadelphia. Officials in Philadelphia allowed a parade to be held in September 1918, and imposed strict social distancing measures 17 days after the first cases were reported. Philadelphia's healthcare

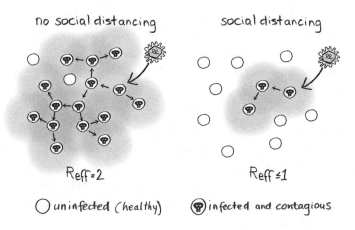

system was quickly overwhelmed, and the number of deaths far exceeded that in St. Louis. In general, data across US cities showed that during the 1918 influenza pandemic, US cities that imposed social distancing policies sooner had smaller peak death rates and smaller overall death rates during the first wave of infections. This is because social distancing reduces human contact, thus making R_{eff} smaller. So, the number of infections and deaths is smaller. This is what epidemiologists mean when they say that social distancing is necessary to "flatten the curve."

The data from the 1918 influenza pandemic showed that imposing social distancing measures early in an epidemic can mitigate the possibility of overwhelming the healthcare system and reduce the number of deaths during the first wave of the epidemic. Many of the measures that were implemented around the world during the COVID-19 pandemic were similar to those that Starkloff put in place in St. Louis in 1918. These measures, which did succeed in flattening the curve in most places, included the following:

1. Encouraging distancing between people, with and without masks
2. Banning of large public gatherings, such as sporting events and concerts
3. Closure of schools
4. Closure of nonessential businesses
5. Travel restrictions
6. Orders to stay at home or shelter in place

Interestingly, data from the 1918 influenza pandemic also show that, when social distancing measures were relaxed after the first wave of infections subsided, the second wave of infections led to higher peak death rates in cities that had imposed social distancing earlier. This was presumably because of a greater proportion of susceptible people in those cities. One important difference between 1918 and the present times is modern medicine. There was little medical treatment that could be provided to sick patients in 1918. During the COVID-19 pandemic, physicians quickly optimized treatment strategies to reduce fatalities. So, after relaxation of social distancing measures and the expected increase in cases, optimized clinical treatment can reduce fatalities even if there is still a larger number of susceptible people in the population.

Later in the chapter, we will return to how mathematical models can be used to suggest ways to mitigate the effects of waves of the epidemic. But first let us ask whether social distancing can completely extinguish an epidemic.

Can Social Distancing "Crush the Curve"?

If strict social distancing is imposed for a long enough time, the value of R_{eff} would drop below 1, and the epidemic would be extinguished. Assume that public health measures were such that no one could leave their homes for any reason. Food would be delivered to the door by government workers who would not have any contact with you, and the same would be true for any drugs needed for chronic conditions. Imagine that at the start of imposing such measures, 5,000 people were infected. With these measures in place, at most they would infect people who shared a home with them, but the cohabitants thus infected would not infect anyone new. So, over time, the R_{eff} would drop below 1, and ultimately the epidemic would wane. This is the situation that is implied by the phrase "crush the curve." Is it possible to achieve this?

Very strictly enforced and intense social distancing measures are said to have crushed the curve in Shenzen and Wuhan in

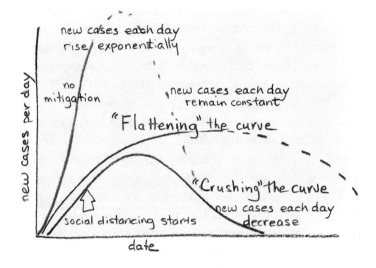

China. After R_{eff} falls below 1, the number of infections starts to fall, but it takes a while for the virus to be extinguished. Crushing the curve requires enforcing strict social distancing measures for a long time and forbidding travel in and out of the region. It seems that crushing the curve is a strategy that can work, but might not be practical in all settings.

Weathering the Storm

Social distancing measures can flatten the curve to save lives during a spreading epidemic and, imposed in extreme form for a sufficiently long time, can extinguish an epidemic. But social distancing measures also have many adverse consequences. The most obvious cost is the economic downturn caused by shutting down businesses and the concomitant loss of jobs. For example, during the first months of the COVID-19 pandemic, 40 percent of US households earning less than $40,000 per year lost their jobs.

Such economic carnage and the loss of citizens' livelihoods can have many ripple effects that include depression and addiction.

Given these serious concerns, and the fact that we do not know exactly how efficacious each of the social distancing strategies noted above are, some have suggested that a better strategy may be to just weather the storm and let the wave of infections pass through. A related strategy, "shield," temporarily shuts down society to allow the isolation of those most at risk for serious illness and death, such as the elderly and the immunocompromised. Restrictions are then relaxed, and the less vulnerable weather the storm. The hope is that people would naturally react to a pandemic by being vigilant, thus not allowing the number of infections and deaths to grow so fast as to overwhelm the healthcare system. Ultimately, the number of infections would die out and things would return to normal. Indeed, before the advent of modern medicine and therapeutics, all infectious disease epidemics ended this way. But why does an epidemic end by weathering the storm?

As the number of people who recover from disease and become immune to infection increases, the proportion of the population that is susceptible to infection drops. To illustrate how this works, let us consider a specific location where the social network is such that an average person interacts with 100 people during an infectious period of 10 days. In the beginning of the epidemic, all 100 people are susceptible to infection. As people become infected and recover, a proportion of the population becomes immune and is no longer susceptible to infection. Suppose we reach a point where 95 percent of the population is immune. Now, only five out of the 100 people an infected person interacts with during the infection could get infected. If the value of R_0 for a particular viral infection at this location is 2,

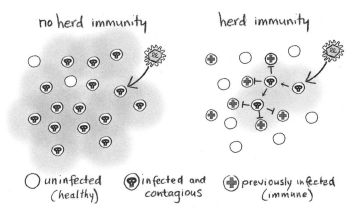

no herd immunity | herd immunity

○ uninfected (healthy)   ⊕ infected and contagious   ⊞ previously infected (immune)

instead of infecting two people, an infected person would only infect $2 \times (5/100)$ people. The R_{eff} in this situation would be far less than 1. So, once a sufficiently large proportion of the population becomes immune, the epidemic ends over time because an infected person is very unlikely to infect anyone else. Even a susceptible person has a very low chance of being infected. When this situation prevails, the population is said to have acquired "herd immunity."

Knowing the proportion of the population that must be immune so that herd immunity is acquired is important. If this proportion is very large, acquiring herd immunity takes a long time because many people have to become infected. If a significant proportion of people infected with the pandemic-causing virus require hospitalization and die, many would die during this time. Let us estimate the proportion of the population that must become immune for a population to acquire herd immunity.

Consider a viral infection characterized by a particular value of R_0, and let us denote the proportion of the population that is immune as p_i. For example, if 50 percent of the population is immune, $p_I = 0.5$. If a person interacts with 100 others during the infectious period, as in the example we considered above, only $100 \times (1 - p_I)$ people that a person encounters are susceptible (50 people). So, instead of infecting R_0 people during the infectious period, an infected individual would infect a proportionately smaller number, which is equal to

$$R_0 \times \frac{100 \times (1 - p_I)}{100} = R_0 \times (1 - p_I).$$

This is the value of R_{eff} now. So, if 50 percent of the population is immune ($p_I = 0.5$), then R_{eff} is half of R_0 ($R_{eff} = R_0 \times 0.5$). Herd immunity is acquired when R_{eff} falls below 1. So, R_{eff} ($R_0 \times (1 - p_I)$) must be less than 1, or, equivalently, p_I must be more than $1 - 1/R_0$. When the proportion of the population who have recovered from the disease exceeds $1 - 1/R_0$, the population acquires herd immunity. So, if R_0 is 2, p_I must be more than one half; that is, herd immunity is acquired when more than 50 percent of the population has recovered from disease.

Because the proportion of the population that must be immune for herd immunity to be established depends on R_0, what it takes to acquire herd immunity depends on the virus and on local conditions. If the R_0 is high, such as 18 for measles, the magic number for herd immunity to be established is $1 - 1/18$, which is about 0.94. So, roughly 94 percent of the population has to be immune. Highly infectious viruses, like measles, are very difficult to control by hoping that natural infection will lead to herd immunity because it requires essentially everyone in the population to have been infected. Vaccines play a key role in such cases by immunizing a large number of people such that herd immunity is achieved without natural infection with the virus.

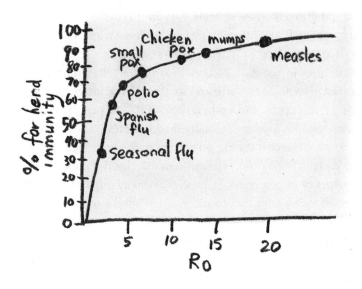

The MMR vaccine, given to children, protects against measles, mumps, and rubella, all viruses with high values of R_0. Almost all states in the United States require this vaccine for preschool and elementary school admission because without herd immunity many children would be infected and some would die. Measles is not a benign virus. It has a mortality rate of about 0.2 percent in the United States, similar to influenza, and infection can also cause permanent neurological damage. Herd immunity acquired by vaccination programs is an important reason for the sharp decline in childhood mortality in the twentieth century. Maintaining herd immunity requires vigilance as outbreaks will occur if the proportion of vaccinated children falls below that required for herd immunity. This is why, for example, measles outbreaks occur frequently in communities where people do not vaccinate their children. When this happens, vulnerable children with compromised immune systems (e.g., due to chemotherapy), who are protected from infection by herd immunity, are at risk again.

If the value of R_0 is not too large, herd immunity can be established by natural infection. In 2016, the Zika virus began to spread from Brazil to North and South America. Infection caused a fever, joint pain, and a skin rash, but up to 80 percent of infections were asymptomatic. Sadly, many babies born to women who were infected during pregnancy had a brain defect called microcephaly. The Zika epidemic was declared a public health emergency by the WHO. Steps were taken by public health officials in many countries to mitigate the spread of Zika, including the development of a vaccine. However, as 2016 drew to a close, new infections started to wane and the epidemic died out. Testing for Zika antibodies showed that in parts of Brazil and El Salvador, hot spots of the epidemic, more than 60 percent of the population was infected. The estimated R_0 for Zika is between 2.1 and 2.5, so herd immunity is acquired when 52 to 60 percent of the population is immune. So, naturally acquired herd immunity may have controlled the epidemic.

For SARS-CoV-2, estimates of the value of R_0 vary between 2 and 3. If the value is 2, herd immunity would be acquired when half of the population is immune. For an R_0 of 3, herd immunity would be acquired when two thirds of the population is immune. During the COVID-19 pandemic, the United Kingdom initially seemed to be on the verge of adopting a "weather the storm" policy. But when mathematical models predicted a scale of infections that would completely overwhelm the healthcare system and cause many deaths, the government reversed course. Sweden, however, decided to essentially weather the storm. They did not lock down the country, nor close schools for younger children, and recommended only personal social distancing measures. By

September 2020, the percentage of deaths in the Swedish population was higher than in neighboring Scandinavian countries, about the same as the overall rate of deaths in the United States, and substantially less than in New York City. Antibody testing suggests that Sweden has not attained herd immunity. Perhaps this is not surprising. It took a few years for the plague to pass through Europe, and more than two years for the 1918 flu to extinguish itself, largely without mitigation. It remains unclear how high the death toll in Sweden will be before herd immunity is acquired naturally or by vaccination. Comparing different strategies that have been used across the world will provide valuable knowledge for mitigating future pandemics.

Many factors, including cultural differences resulting in better compliance with public health measures and self-imposed social distancing, which cannot be anticipated by epidemiological models, may explain why Sweden was able to control infections better than anticipated without a lockdown. Similar factors may have been important for the way that Japan controlled the pandemic. This is also an opportune point to emphasize again that parameters used in models for one location may not be appropriate for another.

During the COVID-19 pandemic, in many countries, such as the United States, social distancing imposed during the first months of the pandemic flattened, but did not crush, the curve. In the United States, as expected, infections did increase as social distancing rules were subsequently relaxed. While, as of September 2020, there is no evidence that herd immunity had been acquired in any country, it may have been achieved in some neighborhoods in New York City. This suggests that in some parts of the world where infection rates were high, the spread of the virus is slowing.

Is a "weather the storm" strategy the only alternative when social distancing measures are relaxed? This strategy could be dangerous. Data from the 1918 influenza pandemic showed that cities that imposed stronger social distancing measures initially experienced a bigger second wave. After the first phase of an epidemic, how does one try to balance the needs of keeping the economy and some semblance of normal life going without overwhelming the healthcare system, which can result in avoidable deaths? Epidemiological models can explore various scenarios and make qualitative comparisons that can guide public officials in this regard.

Acquiring Herd Immunity by Intermittent Social Distancing

One strategy that could be considered for mitigating subsequent waves of infection if the curve was not crushed and herd immunity was not acquired is intermittent imposition of social distancing measures. This strategy aims to keep the economy and normal life going as long as possible without overwhelming the healthcare system. For epidemiological models to explore the consequences of such a strategy and how it should be implemented, the values of various parameters that go into the SEIR models discussed earlier must be known or assumed. For example, one would have to know or estimate the extent to which different social distancing measures influence R_{eff}. The difference in values of R_{eff} with and without imposition of social distancing measures is related to the difference in the parameters that determine the rate at which susceptible people are exposed to the virus ($S + I \rightarrow E$) and subsequently become infected ($E \rightarrow I$). Many other variables are also important. For example, for how long are people who have recovered from the disease immune? What is the seasonal variation in R_{eff}? Is it larger in the winter, as it is for influenza? With clearly stated assumptions, which are important to know, epidemiological models can offer some useful insights.

As just one example of such an insight, we describe projections made in April 2020 by Grad, Lipsitch, and colleagues (epidemiologists at Harvard University) about how intermittent social distancing might affect the COVID-19 pandemic under different scenarios. These epidemiologists assumed that R_{eff} cycled between higher values in the winter and smaller values in the summer. The trigger for imposing social distancing measures should be when new cases start to rise and exceed a threshold value. This threshold value would be set to prevent overwhelming the capacity of the healthcare system. When the number of cases declines below this threshold, social distancing measures would be relaxed. As time ensued, more and more people would be infected and become immune, and so the population would progress toward herd immunity. An interesting effect of healthcare capacity was predicted. If hospital capacity was increased, the threshold value of cases when social distancing measures are turned on could be higher. This is because the healthcare system would be less easily overwhelmed. Thus, periods with no social distancing measures could be longer. This would enable the population to acquire herd immunity faster, and keep normal life and the economy going longer. As more people became immune, the duration between imposition of social distancing measures would get longer as R_{eff} would be smaller and fewer people would get infected. Ultimately, when the number of new cases becomes small enough, testing, isolation, and contact tracing will be sufficient to control the epidemic.

These types of projections are very useful for guiding public policy. But without real-world data, the predictions cannot be quantitatively accurate. For example, to make models like this

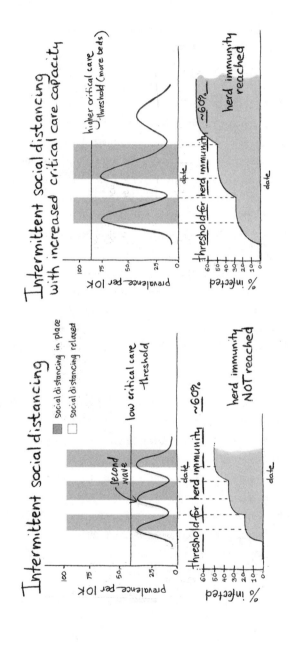

predictive, we need extensive testing to know what proportion of the population is immune. We need immunological studies to determine whether immunity declines and how long it lasts. Many other parameters are also needed, some of which can be difficult to measure.

Massive amounts of data are being acquired across the world about the COVID-19 pandemic. At the same time, we have enormous computational resources today. Modern machine-learning approaches could mine the data being collected to obtain an understanding of pandemic control that far exceeds anything we have had before. Perhaps, in the future, this understanding will allow us to develop models that can help find public policies that optimally control hospitalizations and deaths to acceptable levels while not causing other kinds of damage to society. Thus, we can hope to be more resilient when the next pandemic comes.

The quickest way to end a pandemic that is difficult to control is to have an effective drug that cures the disease, or a vaccine that provides herd immunity. We now turn to these two topics.

6 Antiviral Therapies

In 1981, five young gay men in the United States developed pneumonia caused by a fungus called *Pneumocystis jirovecii*. Previously, pneumocystis pneumonia was an extremely rare disease that developed mainly in immunocompromised people. The cluster of cases affecting otherwise healthy young men suggested that their immune systems were compromised, thus reducing their ability to fight off infections like pneumocystis pneumonia. This was because a new virus, HIV, that infected immune cells had entered the human population.

Almost 40 years later, we still cannot cure HIV, but therapeutics have been developed that allow HIV-infected people to live relatively normal lives. In 1995, a combination of antiviral drugs was developed that can keep the amount of virus in an infected person at very low levels. Before these drugs became available, a positive test for HIV was essentially a death sentence.

The approaches used to develop antiviral treatments for HIV forged a path that has led to successful development of treatments for other viral infections. In this chapter, we will explore the strategies that are used to develop drugs that can treat viral infections.

Identifying the Virus

The first step toward developing therapies that can treat a disease caused by a new virus is to identify the causative virus. This is because antiviral drugs interfere with the virus's ability to replicate in humans and cause disease. We have to know what type of virus it is in order to know how it replicates, and so how to interfere with these mechanisms. Identifying a virus was a formidable task until the middle of the twentieth century because they are so tiny. In 1949, John Enders, Frederick Robbins, and Thomas Weller transformed the study of viruses when they figured out how to make poliovirus multiply and grow in animal cells in test tubes. Enders and colleagues used these methods to grow, identify, and study viruses, including those that cause measles and mumps. This work directly led to vaccines that protect against the childhood diseases caused by these viruses. Enders, Robbins, and Weller would be awarded a Nobel Prize for their achievements. Others quickly adopted their approach to growing a virus in the laboratory making it much easier to identify and study viruses.

Françoise Barré-Sinoussi and Luc Montagnier in France were the first to identify HIV, followed by Robert Gallo in the United States. Their key insight was that the virus could be grown in a subset of T cells, the same cells that the virus infects to cause disease. They knew that they had identified the disease-causing virus because the tests they developed showed that the virus they grew in the laboratory was the same as the one present in all infected patients. HIV was identified about two years after the first reports of a new disease. This was a stunning achievement because it took decades to identify disease-causing viruses like polio.

Modern technologies have revolutionized the speed with which a virus can be identified. In 2003, about 6 months after reports of the disease, the virus that caused SARS was identified. In 2020, the genomic sequence of SARS-CoV-2, the virus that causes COVID-19, was identified about a month after the first reports of the disease. Quick identification of new viruses is made possible by new methods for rapid isolation of viruses and fast methods to sequence their genomes.

The Life Cycle of Viruses Defines Targets for Antiviral Drugs

In chapter 3, we learned how viruses function. Let us briefly review some aspects that are relevant for the development of antiviral therapies. Viruses have to replicate rapidly so that they can infect many cells in a person and many people in a population. Viruses have only the proteins that are absolutely essential for their functions. They do not even have most of the proteins that they need to replicate, which means that to multiply they need to hijack proteins from the cells they infect. Viruses can multiply rapidly because only a few proteins need to be replicated to form new virus particles. Viruses are like experienced travelers who pack very light, carrying only items that they absolutely need and that are not provided by hotels. Because viruses have only the proteins that are essential for their function, any drug that can inhibit the function of any viral protein is a potential antiviral drug.

All viruses have a similar life cycle. Viruses enter the body through the skin, eyes, mouth, nose, rectum, vagina, or other vulnerable body part. Insect bites or hypodermic needles allow viruses to directly enter the bloodstream. Once inside the body, they enter our cells. Entry into a cell is mediated by the virus's spike binding to a receptor on the surface of our cells. For example, the HIV spike binds to a receptor on the surface of a subtype of T cells, and SARS-CoV-2 binds to the ACE2 receptor present abundantly on lung, heart, and kidney cells. Once the virus's spike is attached to a cell's receptor, the next step is to force its way into the cell. This is a complex step that often involves our own cell's proteins and a change in shape of the proteins that make up the virus spike. Once inside the cell, the viral genome is released. Our own cell's machinery is then hijacked and the cell becomes a factory for replicating the virus's genome to make many copies of its few proteins. The proteins are then assembled into many new virus particles. At this point, the newly assembled viruses are still trapped in the cell and need a way out. Viruses have developed many ways to escape from the cell.

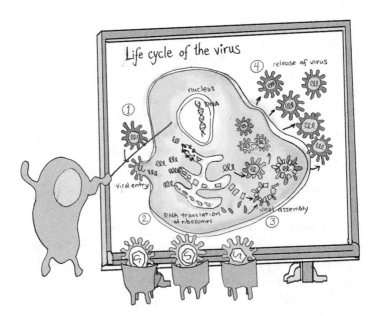

Some viruses rupture the cell in order to exit, while others can just bud out of the cell's soft membranous wall. Once outside the cell, the newly produced viruses infect other cells, and the cycle is repeated until the infected person dies or the immune system controls the virus.

Antiviral therapies aim to block one or more of the steps in the viral lifecycle: viral entry, replication, assembly, and release from the cell.

Blocking Viral Entry

Blocking viral entry into the cell is a proven antiviral strategy. In fact, this is a strategy that our immune system uses effectively. Some antibodies generated in response to the virus attach to the virus's spike and block its ability to bind to receptors on human cells. This prevents the virus from infecting new cells. This is why for over a hundred years antibodies have been used to treat disease.

As we described in chapter 4, von Behring and Kitasato showed that antibodies injected into patients could treat diphtheria. Later, physicians used blood (plasma) from patients who recovered from disease to treat viral infections that include influenza, measles, SARS, MERS and Ebola. The idea that antibodies in blood can be curative is why plasma from patients who have recovered from COVID-19 is being tested as a therapy.

While many antibodies are generated in response to the virus, only some are really potent in preventing viral entry. Identifying the potent antibodies and using these as drugs can be an effective form of therapy. Many companies and scientists have developed rapid ways to generate these desired antibodies. For example, a clinical trial for Ebola virus infection tested a combination of

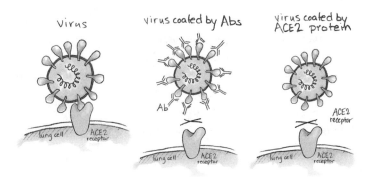

three such antibodies specific to Ebola that were produced by the biotech company Regeneron. The trial was so successful that it was stopped early. Similar approaches are being used to identify effective antibodies for the treatment of COVID-19. These therapies induce a transient immunity. Questions about whether only one or multiple types of antibodies are required, and which particular antibody types (IgG or IgA) are needed remain unclear.

Finding the right antibody can be a challenge, as it can be a bit like looking for a needle in a haystack. A way to take a short cut is to use the cell receptor that a virus's spike attaches to as the "antibody." For example, ACE2 is the cell surface receptor that the SARS-CoV-2 virus binds to enter the cell. One strategy might be to manufacture ACE2 and use it as a decoy. Once injected into a patient, the synthetic ACE2 receptor would bind to the virus and prevent it from binding to the ACE2 receptor on the cell. The synthetic ACE2 receptor drug could also be engineered to have one end that is like the stem of an IgG antibody. Then the mechanism that helps the body dispose of antibody-bound viruses could get rid of the virus–drug complex.

Antibodies and receptor decoys are large molecules, in a class of drugs called biologics. These types of drugs are expensive, partly because they are more complicated to manufacture compared with drugs that are small molecules. They require injection or intravenous administration, making them difficult to deploy in large scale. Their use is probably more appropriate for treating the severely ill, either to suppress geographically localized outbreaks or to protect family members of an exposed individual. Administration of antibody therapy to everybody in a neighborhood with an outbreak would, for example, temporarily provide immunity for the community and suppress spread of the infection. A single dose would likely be sufficient since injected antibodies persist in the blood for some time.

Once bound to a receptor on the cell, viruses need a way to force their way in and enter the cell. Viruses have figured out many ways to do this. In general, upon binding to the receptor on the cell surface, the viral spike protein can dramatically change its shape, which via complicated mechanisms generates a force that allows the virus to push its way in. Blocking this step with a drug can prevent the virus from entering the cell even if it is attached to the appropriate cell surface receptor.

SARS-CoV-2 uses a protein already present on the surface of lung cells, called a protease, to cut the viral spike protein into two pieces. Cutting the viral spike protein acts like a spring being released, and the resulting force allows the virus to push its way

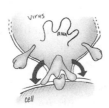
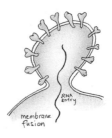

in through the cell wall. A drug that could block the action of the protease that cuts the spike could prevent viral entry into the cell. Since the protease is present on many cells in different parts of the body, a drug blocking it could have wide-spread and serious adverse side effects. Preventing such side effects is a major challenge for drug development.

Blocking Viral Replication

As noted above, viruses have only the proteins that are essential for their function, and which they cannot steal from our cells. As we saw in chapter 3, RNA viruses have an RNA genome. So, they cannot make their proteins following the DNA to RNA to protein route that we use (recall the central dogma). So, all RNA viruses have their own polymerase, which allows them to copy their genome and replicate. Some DNA viruses also have their own polymerase that copies their genome. For retroviruses, like HIV, the polymerase is called reverse transcriptase. It converts the viral RNA genome to DNA, which is then converted to RNA and proteins using our cell's machinery. A drug that could specifically inhibit the viral polymerase would be efficacious and safe. It would be efficacious because it would prevent viral replication, and safe because the ideal drug would not act on our polymerase, which is distinct from that of the virus. The trick is how to find such a drug.

A solution was provided by Gertrude Elion in the late 1970s. After graduating from high school at the age of 15, Elion attended Hunter College of the City College of New York. Despite graduating with top honors in chemistry in 1937, because of gender discrimination, her applications to graduate school were rejected multiple times. She could only find employment as a secretary and then as a food quality supervisor at a grocery store chain before finally finding a job as an assistant to George Hitchings at the Burroughs Wellcome pharmaceutical company. Hitchings was developing antibiotics, and he thought that blocking an organism from replicating its DNA might be a good strategy to stop it from multiplying. Recall from chapter 3 that DNA and RNA are chains of units called bases. These bases are G, T, A, and C for DNA and G, U, A, and C for RNA. Polymerases take these units and string them together in the specific order that encodes the genome of a particular cell or organism. Elion and Hitchings synthesized molecules that looked very similar to the usual bases. They are similar enough that the polymerase is fooled into inserting it into a growing genome. Using these molecules, they succeeded in developing antibiotics and drugs for cancer and autoimmunity.

After Hitchings retired, Elion continued their work. In the late 1970s, she began to work on developing a drug that could specifically inhibit the replication of the herpes virus (a DNA virus). Her strategy was to synthesize artificial bases that could be inserted by the viral polymerase into a growing copy of the viral genome. Since these drugs are not true bases, once inserted, they block the polymerase from growing the chain any further. Thus, the genome stops growing before it is completely copied. This defective genome no longer encodes information about all the necessary proteins, and so the virus cannot replicate. Using this strategy, she identified the first antiviral drug, acyclovir, which the Herpes virus's polymerase acts on, but our own polymerase does not. Early in the HIV pandemic, her team used this strategy to identify the first drug used to treat HIV, azidothymidine (AZT, zidovudine). Remdesivir, a drug being used to treat COVID-19 is also a polymerase inhibitor. It was originally designed to inhibit

Inhibition of viral polymerases by drugs

without drug

with drug
chain termination

polymerase inhibiter = Z

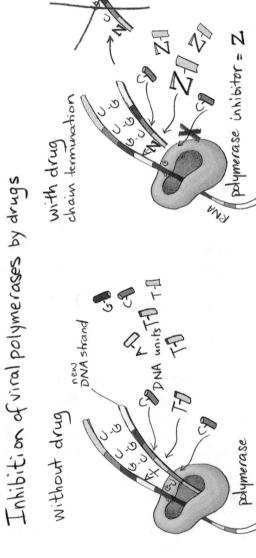

the Ebola virus's polymerase but was found to also have activity against the polymerase of SARS-COV-2.

Elion, who was not allowed to go to graduate school, shared the 1988 Nobel Prize with Hitchings for their work. Only after this would she be awarded honorary doctorates from NYU and Harvard, and elected a member of numerous honorary scientific societies.

For HIV, we now have newer strategies to block its replication. To be as compact as possible, HIV has evolved to have only a few genes, each of which encodes information on multiple proteins. Large polyproteins, containing many individual proteins, are first made. The polyproteins are then cut into the individual proteins that HIV needs to function. Like the protease on our lung cells cuts the spike of SARS-CoV-2 to enable it to enter our cells, HIV has its own protease, which cuts its polyproteins into the right constituent proteins. Without this protease, HIV's proteins, including its polymerase, cannot be separated from the corresponding polyprotein, and so would not function. Drugs called protease inhibitors were developed to block the function of the HIV protease, and this prevents virus replication.

Blocking Assembly of the Virus

Once the virus's component proteins have been made in the infected cell, these parts need to be assembled into complete viruses. This is another step in the viral lifecycle that can be interrupted by drugs.

Hepatitis C virus (Hep C) is transmitted through the blood and causes a liver infection. The immune system is unable to eradicate this virus in most patients. Thus, if untreated, Hep C causes a chronic, lifelong infection in most people. About 2–3 percent of the world's population is thought to be infected with Hep C. Patients are generally asymptomatic during the early stages after infection, but the virus slowly destroys the liver. Hep C is now one of the most common causes of liver failure.

Until the virus was identified about 30 years ago, it was a mysterious infectious agent that contaminated the blood supply. Receiving a blood transfusion was a bit like Russian roulette, and transfusions were a relatively common way to contract the virus. The identification of Hep C was a key moment because not only did it provide a way to screen for the virus and clean up the blood supply, but it also allowed the search for effective treatments.

The first antiviral drugs used to treat Hep C had been developed in the 1970s and 1980s. One of these is called ribavirin. Ribavirin is similar to the drugs developed by Elion and Hitchings in that it resembles one of the bases that make up a virus's RNA genome. When it is added to the growing RNA genome as it is being copied, it has several effects that include slowing chain growth. The base-like drug is also such that wrong bases can add to it. This introduces so many mutations into the copied RNA that the resulting virus is dysfunctional.

The second drug developed to treat Hep C was interferon. Interferon is a hormone made by immune cells that creates an environment that is very inhospitable for viruses. This drug was made using recombinant DNA technology, which was invented in the 1970s by Paul Berg, Stanley Cohen, and Herbert Boyer. This method allowed the gene for interferon to be identified and isolated in the laboratory. This gene was then inserted into the DNA of another organism, which then produced "synthetic" interferon encoded by the inserted gene. This synthetic product was administered as an anti-viral treatment. Producing interferon in this way was a landmark event for the biotechnology industry. But the combination of ribavirin and interferon that was used to treat Hep C infections was effective in only about 50 percent of patients.

To improve upon antiviral drugs like ribavirin and interferon, the search for Hep C–specific drugs began. Drugs that specifically inhibited Hep C's polymerase and protease were developed, and they significantly increased the cure rates for the disease and reduced the need for liver transplants later. These drugs shortened the treatment period for Hep C patients from about a year to only six weeks and cured the disease in a significant proportion of patients. These drugs are convenient, requiring patients only to swallow pills. But the initial cost of the treatment, over $1,000 per pill (for Solvadi), stirred great controversy.

An important breakthrough in the treatment for Hep C was the development of drugs that block the assembly of the virus. These drugs inhibit the action of a Hep C protein called NS5A. We do not know exactly what the function of NS5A is, but the use of NS5A inhibitors in combination with protease or polymerase inhibitors is one of the most effective antiviral therapies ever developed.

Blocking Release of the Virus

Once new viruses are assembled, they need to exit the cell. This is another step that drugs can block. Two influenza drugs, oseltamivir (Tamiflu) and zanamivir (Relenza), interfere with the release of viruses from the cell. As you recall, viruses need to bind to a receptor to enter the cell. But if a new virus particle binds to its receptor before exiting the cell, it will get trapped in the cell. After new viruses are assembled, the influenza virus deploys a protein that destroys its receptor, allowing newly assembled viruses to slip out of the cell without getting stuck. Oseltamivir and zanamivir block this function, thus trapping the virus in the cell. Since these drugs block only the last step of replication, they are less effective than other types of drugs. This is because the virus has already assembled many copies of itself and some can still escape from the cell. Oseltamivir and zanamivir just shorten the length of the illness by one or two days.

Combination Therapies

Like a stunt where Houdini miraculously escapes from a strait-jacket, viruses have an uncanny ability to develop resistance to drugs. Recall from chapter 3 that the replication of viral polymerases is relatively inaccurate, and every time the virus replicates, mutations can arise in its genome. After a few cycles of replication, different strains of the same virus may emerge. When confronted with a drug, viral strains that have a mutation that allows them to function without being impaired by the drug will have an advantage over those that are affected by the drug. These mutant strains then thrive, and the drug is no longer effective. The development of drug resistance by this mechanism is very common for highly mutable viruses like HIV.

Pharmaceutical companies have developed drugs that inhibit the functions of distinct proteins of a virus. If a combination of these drugs is used simultaneously, a drug-resistant viral strain would need to acquire multiple mutations at the specific targets of all these drugs. This is harder to do because viral polymerases make mistakes randomly, and so mutations arise in viral genomes at random spots. The emergence of multiple specific mutations in a viral strain by this random process takes a while. It's like if the virus was playing roulette, to win against one drug it would have to guess one number correctly, while to win against a drug cocktail it would have to guess multiple numbers correctly to win. Moreover, each mutation likely results in some impairment of the virus's functions—many would make the virus even more dysfunctional. For these reasons, combination drug therapy has been very successful for treating highly mutable viruses. For example, modern HIV and Hep C treatments typically employ a cocktail of drugs.

Coping with the COVID-19 Pandemic

At the beginning of a pandemic caused by a novel virus, there is not enough time to develop new drugs specific to this virus. Trying to save patients' lives, doctors repurpose drugs approved for other diseases or viral infections, hoping that they will work. This is a good strategy since these drugs are known to be safe, and their side effects are known. During the COVID-19 pandemic many such therapeutic strategies were tested.

The pharmaceutical company Gilead has expertise in viral polymerase inhibitors, having successfully developed such drugs for Hep C. They had previously repurposed a polymerase inhibitor that was originally developed for Hep C to treat Ebola infections. This drug, remdesivir, had been tested in humans in Africa and was established to be a safe and promising therapeutic. However, when Regeneron's antibody treatment for Ebola proved to be more efficacious, remdesivir's development was halted before it was approved for use.

In 2020, Gilead quickly tested remdesivir for efficacy in inhibiting the polymerase of SARS-CoV-2. It was found to inhibit virus replication both in animals and in human cells in test tubes.

Gilead rushed to perform clinical trials with the drug and also began to provide it on a compassionate use basis. When the drug appeared to shorten the course of disease, the US Food and Drug Administration granted Emergency Use Authorization for treating severely ill patients. The drug requires intravenous administration and causes some side effects, limiting its use to the severely ill.

Dexamethasone is a steroid that has long been used to treat arthritis, acute respiratory distress, and other conditions. A British clinical trial has shown that use of this drug reduces the death rate of COVID-19 patients who were on ventilators by a third, and by 20 percent for those on external oxygen support. This is a good example of a drug that has been repurposed for treating acutely ill patients during a pandemic.

As we described earlier, another approach being pursued to treat COVID-19, is taking plasma from those who have recovered from the infection and injecting it into sick patients. As each individual is different and could mount a different kind of antibody response, there could be high variability in the efficacy of antibodies obtained from different donors. Several companies are developing potent neutralizing antibodies that can be administered to patients, which will eliminate this complication.

Two types of interferons are approved as drugs for the treatment of Hepatitis B and Hep C infections, as well as for the autoimmune disease multiple sclerosis. There are data to suggest that SARS-CoV-2 has a gene that functions specifically to block the production of interferon, so it may be a good strategy to consider interferon treatment. The usefulness of the two approved forms of interferon for COVID-19 has yet to be established. Also, these drugs are expensive, need to be injected into patients, and have significant side effects, which hinders their use at large scale.

Quinine is a medication isolated from a tree bark that has been used for the treatment of malaria for almost 400 years. Because of its bitter taste, the British mixed it with gin to create the popular drink known as gin and tonic. Great efforts were made in the 1930s to make analogs of quinine that could be synthesized in the laboratory. This led to the discovery of chloroquine and, later, hydroxychloroquine by chemists at the German company Bayer. At the end of World War II, US troops captured the drug, and in 1947 it was found to be effective in preventing transmission of malaria. The drug was also repurposed to treat some autoimmune diseases. Although it has been used for so long, we still do not really understand how this drug works. The excitement about it possibly being able to treat COVID-19 reflects the paucity of available treatments. Only carefully controlled clinical trials can determine whether drugs might be effective in controlling the virus. Current data suggest that hydroxychloroquine is not an effective treatment for COVID-19.

Inhibiting a Cytokine Storm

As we described in chapter 4, sometimes an overexuberant immune system can cause severe disease. This can arise due to either extensive spread of the virus, which elicits too strong an immune response, or a faulty immune system. An overly active immune response can lead to the production of large amounts of cytokines, a so-called cytokine storm. Dramatic effects result, including acute inflammation, a precipitous drop in blood pressure, and uncontrolled blood clotting throughout the body. The major cytokines that are secreted during a cytokine storm are called IL6, TNF, and IL1. Drugs that block the action of these cytokines are already approved for the treatment of a variety of autoimmune diseases. If cytokine storms are the cause of death in the most severely ill patients, identifying which patients could benefit from cytokine blockade and which cytokines are important to block could play a role in reducing the mortality of COVID-19.

The Future of Antiviral Drugs

Developing efficacious antiviral drugs is a time-consuming and expensive proposition. Identifying a good drug candidate is very hard. But this is just one step in the process, as expensive clinical trials are needed to establish that a drug candidate is safe and efficacious. This, in turn, requires optimizing the dose to be used and the method of administering the drug. The majority of drugs that are developed fail to get approved. For infectious diseases that are likely to continue to afflict us for a while, such as those caused by HIV, Hep B, Ebola, and influenza, companies continue to devote effort and expense to develop good therapies. To prepare adequately for potential pandemics, we will need a different model for developing antiviral drugs. These models will be predicated on new scientific advances and approaches, and will be facilitated by public-private partnerships. We will opine on these issues in the epilogue.

The most effective strategy to end the scourge of an infectious disease– causing virus is to develop a vaccine that protects against it. That is the topic of the next chapter.

7 Vaccines

Human history is inextricably linked with the pain inflicted on us by infectious disease-causing microbes. Until relatively recently, most families would lose at least one child to infectious diseases. Smallpox killed hundreds of millions of people in the twentieth century alone. In present times, especially in the developed world, it is hard to imagine this kind of trauma so common even 100 years ago. A large part of the reason that this situation has changed is the development of vaccines that can protect us from infection and disease. Indeed, vaccination has saved more lives than any other medical procedure. Comprehensive smallpox vaccination programs have saved hundreds of millions of lives and led to its eradication from the planet. Vaccination has also almost eradicated polio and is a major reason for the dramatic decline in childhood mortality.

As we discussed in the chapter on pandemic mitigation, infectious disease–causing viruses stop spreading in a population when the proportion of the population that becomes immune to the disease rises above a threshold. For highly infectious viruses, this threshold is very high. So, it is difficult to acquire herd immunity naturally without many people being infected, which can take a long time and potentially cause many deaths. Vaccination allows a population to acquire herd immunity rapidly. Vaccination protects not just the immunized person but the whole community, including the most vulnerable members of society, such as the elderly and immunocompromised.

We began this book by describing Edward Jenner's invention of a safe smallpox vaccine. In this chapter, we will describe how vaccination has evolved since the time of Jenner and Pasteur, and how modern technologies are addressing the challenge of producing a COVID-19 vaccine.

How Vaccines Work

In the chapter on immunity you saw that, upon infection, the immune system mounts a multipronged response aimed at eradicating the virus from the body. The first responders are the cells of the innate immune system, which try to control virus replication and fence in the virus near the site of infection. These cells secrete cytokines that make the environment inhospitable for the virus. Phagocytic cells also eat up the virus and carry them to lymph nodes. Here, B and T cells of the adaptive immune system interact with the virus's proteins, and subsequently an immune response tailored to the infecting virus is produced. There are two arms of this response. One arm is comprised of antibodies that bind to the spike proteins of the virus and thus prevent the virus from entering our cells and infecting them. The second arm is comprised of T cells. Killer T cells detect specific fragments of the virus's proteins that bind to our HLA proteins. These HLA-bound protein fragments are displayed on infected cells. Upon detecting a viral protein fragment, killer T cells secrete products that punch holes in the infected cells and kill them. We mentioned in the chapter on immunity that there are other subtypes of T cells that have diverse functions. One of these subtypes, called a "helper" T cell, helps mediate the Darwinian evolutionary process that leads to the production of potent antibodies. After an infection is cleared, antibodies and memory T cells and B cells specific for the infecting virus circulate in the body. These products of adaptive immunity can mount rapid and robust responses when needed, thus protecting against reinfection, at least for some time.

The goal of vaccination is to stimulate the immune system to produce antibodies and memory T cells and B cells that attack the specific virus from which we wish to protect the population. Moreover, vaccines aim to stimulate antibody and memory immune responses that are durable for a long time, ideally for the lifetime of the individual. All of this has to be achieved in a way that is safe so that vaccination does not result in a full-blown form of the disease or lead to any other adverse effects. While minimizing, or preferably eliminating, side effects is an important consideration for the development of any therapeutic, this is especially important for prophylactic vaccines as they are administered to healthy people.

The vaccine must contain the virus's proteins, either in whole or in part. Otherwise antibody responses specific to its spike protein or T cell responses to fragments of this specific virus's proteins will not be elicited. All effective prophylactic vaccines induce potent antibody responses. Antibodies prevent the virus from infecting cells, which in turn prevents the virus from multiplying. Vaccines may also need to elicit T cell responses. One of the most successful vaccines protects against infection by the virus that causes yellow fever. This virus was a major health hazard for decades. The deaths of French soldiers due to yellow fever

was one of the reasons France sold their North American hold-
ings to the United States in 1803 (the Louisiana Purchase). The
yellow fever vaccine developed in the 1930s elicits antibodies
and potent and long-lived killer T cell responses. Yellow fever is
still endemic in parts of Africa and South America.

From the chapter on immunity, you may recall Janeway
talking about the immunologists' "dirty little secret." That is,
without a proper innate immune response, there is no adaptive
immunity. This means that an effective vaccine must also induce
a potent innate immune response. We do not understand innate
immunity as well as we do adaptive immunity. Even for the latter, we do not really know what factors lead to durable memory
responses and antibodies that offer protection for a long time.
It is also important that memory T cells are localized to, or can
quickly get to, portals in our body through which a particular
virus enters us, as this would quickly terminate infection. We do
not know much about how to achieve this goal either. So, vaccine development remains a somewhat empirical process. Many
approaches are tried, and some strategies work better than others
in different applications. Let's discuss some of these approaches
to vaccine design.

Types of Vaccines

Live Attenuated Vaccines

Live attenuated vaccines are composed of the real virus that
causes the disease. But the virus used in the vaccine is one that
has been weakened (attenuated) so that it does not cause the
full-blown disease. When people are immunized with these vaccines, the virus in the vaccine infects cells and multiplies, thus
generating an immune response.

The smallpox vaccines described in chapter 1 were "live" vaccines. In the process of variolation, the pus that was collected from patients contained the live virus. You may remember that this material was dried and stored for a while before it was administered by an "experienced person." The drying and storage likely damaged the virus sufficiently so as to usually not cause full-blown disease. The "experienced person" probably knew how much of this material to administer so that it was safe. But variolation was a dangerous procedure and sometimes caused smallpox outbreaks. Jenner also used a live virus, but it was the cowpox virus. Viruses that infect animals usually do not thrive in humans because these viruses have adapted to infecting and multiplying in animals, and we are different. Jenner's vaccine was safe because it was a virus from an animal. It was efficacious because the proteins of cowpox and smallpox viruses are similar, and so antibody and T cell responses that developed upon vaccination with the cowpox virus were also specific for parts of the smallpox virus.

Recall that Pasteur serendipitously made a big advance toward the creation of safe live attenuated vaccines with his chicken cholera vaccine. He then took this a step further with his rabies vaccine. Animal viruses that are not adapted to multiply well in us can jump to humans when they acquire the right mutations or when the right kinds of recombination of different viruses occur. While it is unclear whether they knew anything about all this, to weaken the virus, Pasteur and his collaborator, Roux, turned this fact on its head. They took the virus that afflicted dogs and humans, and repeatedly infected rabbits with it. Sequentially passaging the virus through rabbits in this manner likely led to the virus adapting to multiply better in rabbits, making it less lethal for humans. The rabies virus infects the spinal cord; to

Virus-based vaccines

Live attenuated vaccines

Inactivated vaccines

weaken the virus further, Pasteur and Roux air-dried spinal cords from infected rabbits and used the strains thus derived in their vaccine.

Human viruses can also be weakened by growing them in cells in the laboratory. As mutations emerge, the ones that multiply best in these cells take over the population. These viruses are likely to not multiply as well in humans because they are no longer optimized to thrive in an environment that includes human immunity. Thus, they may be suitable for use as live attenuated vaccines. Different approaches continue to be developed to attenuate viruses so that they can be used as vaccines.

Weakened or live attenuated viral vaccines, while effective, can still sometimes cause the full-blown disease associated with natural infection. This risk motivated the search for vaccines that are not composed of a live replicating virus.

Killed or Inactivated Vaccines

In the late nineteenth century, two groups of scientists, in the United States and in France, showed that killed bacteria could be effective vaccines. The killed or inactivated microbes cannot replicate in humans anymore, and therefore they cannot cause disease. This is accomplished by treating the microbes with various chemicals or subjecting them to high temperatures. One chemical that has been used for this purpose is formaldehyde, which is used for embalming the deceased. Inactivating a microbe using such a chemical has to be done carefully. Using too little formaldehyde will not kill the microbe, and using too much will overly damage it. An intact microbe is required because otherwise vaccination with it will not elicit immune responses to the microbe's proteins in the way in which they are displayed on the real virus. For example, if the inactivation procedure alters the way that the spike protein is displayed, the vaccine will not be able to elicit antibodies that bind to the spike protein of the real virus. Thus, the vaccine-induced antibodies would not be effective in preventing infection.

For reasons that are not entirely clear, inactivated vaccines elicit weaker immune responses than live vaccines. So, inactivated virus vaccines often require a "booster shot." By administering the same vaccine again, the memory B cells and T cells generated during the first vaccination get reactivated and multiply. The potency of the antibodies can also increase.

Subunit Vaccines

In the chapter on antiviral therapies, we described how recombinant DNA technology created the modern biotechnology industry. The gene encoding information about a single protein could be isolated and then inserted into another DNA fragment, and the protein could be produced by growing it in cells. With the advent of these methods, scientists started thinking about how this technology could be used to make safer vaccines.

Since antibodies and T cells target only parts of viral proteins, it seemed logical to use recombinant DNA technology to produce a few viral proteins in large quantities and use these proteins as the vaccine. These so-called subunit vaccines would obviously be safe, as they do not contain the virus at all, just some of its proteins. These vaccines would also be much simpler and easier to manufacture compared with the laborious processes associated with growing viruses or inactivating them just right.

A vaccine that protects against hepatitis B infection was one of the first ones made using recombinant DNA technology. Hepatitis B is a virus that infects the liver and is a major cause of liver cancer and liver failure around the world. Most people's immune systems can clear the infection, but some cannot, and they become chronically infected. About 3.5 percent of the world's population is living with chronic hepatitis B infection. Chronically infected persons can spread the virus to others who come in contact with their blood. Infants born to infected mothers can also be infected.

The spike protein of the hepatitis B virus could be mass-produced using recombinant DNA technology. But the protein was not the only component of the vaccine. Remember the immunologists' dirty little secret: if you inject a foreign protein by itself, there is no adaptive immune response. This is because you need to add other components to activate the innate immune system. These additional chemical components added to vaccines are called adjuvants.

Until we learned about the receptors of the innate immune system (see chapter 4), adjuvants were mostly formulated using trial and error. Some of the early adjuvants were killed bacteria. Immunologists did not know it at the time, but killed bacteria stimulate certain innate immune receptors. Another commonly used adjuvant in many vaccines is aluminum salt. Exactly how it works to stimulate the immune system is not known, but it may function as an irritant that causes tissue damage and inflammation, which activates innate immunity. In the modern era, with our knowledge of innate immune receptors, drugs that specifically target specific innate receptors can be used as adjuvants. But while this new knowledge serves as a guide, adjuvant formulation remains empirical.

Another example of a successful vaccine made using recombinant DNA technology is the one that protects against human papilloma virus (HPV) infection. HPV is an important cause of cervical cancer in women, and protecting against infection by this virus prevents the cancer. The spike protein of HPV produced using recombinant DNA technology aggregates to form a structure that resembles the array of spikes on the virus. This results in the generation of a strong protective antibody response.

Subunit vaccines are also designed using another approach. A virus that can infect human cells is chosen. For example, one could choose a member of the adenovirus family of viruses, which cause mild colds and diarrhea. The genes of this virus are

Subunit vaccines

Recombinant protein vaccines   Viral vector vaccines

purified protein

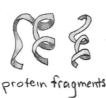

protein fragments

modified so that it cannot replicate after it infects a human cell, and so it cannot cause illness. Using recombinant DNA technology, the genes encoding information on key proteins of a virus for which a vaccine is needed are then inserted into the DNA of the modified adenovirus. The modified adenovirus is called the vector and the inserted genes of the virus is called the insert. When this vaccine is injected into the body, the vector enters human cells and the inserted genes enable production of the corresponding proteins. Immune responses are thus elicited against these proteins. Vaccines designed in this way do not need an adjuvant, because the vector is a live virus that is able to induce innate immune responses. This approach is being used for some of the COVID-19 vaccine candidates.

DNA and RNA Vaccines

Developing and deploying a process for making the subunit vaccines described above takes a long time, often years or more. DNA is transcribed into RNA, which is then translated into the corresponding protein. Technology for synthesizing DNA and RNA has advanced to the point where these molecules can be manufactured in a matter of days to weeks. So, it may be easier and faster to just inject DNA or RNA into people. If the DNA or RNA can enter human cells, the cells would then make the corresponding proteins, which would elicit an immune response. This is a new vaccination strategy that is currently being tested. The first COVID-19 vaccine that entered into clinical trials used such an approach.

The technology for delivering RNA and DNA vaccines in a way that enables them to enter cells efficiently is still evolving. For RNA vaccines, in current methods, the RNA encoding information about the protein of interest is encapsulated in a tiny particle. These particles are called nanoparticles because they are a few nanometers in diameter, which is a thousand times smaller than the diameter of a human hair. The nanoparticles are made up of lipids and other substances, which are the same molecules that make up our own cells' membranous walls and that of viruses. They are designed so that they are preferentially eaten up by Metchnikoff's phagocytic cells, but they can also enter other cells. Once inside the cell, the RNA is then translated into the corresponding proteins, which can then potentially elicit immune responses.

RNA vaccines have the advantage that an adjuvant is not needed. RNA is usually present only inside the cell. When innate immune receptors detect the presence of RNA outside cells, it is like a foreign invader and an innate immune response is activated. The "secret sauce" that makes some RNA vaccines purportedly work well is how the nanoparticle is formulated and how the RNA is modified so as to not induce too strong an innate immune response.

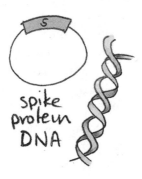

RNA vaccines

spike protein RNA in a lipid droplet

DNA vaccines

spike protein DNA

DNA is much more stable than RNA, and so does not need to be encapsulated in a nanoparticle before it is administered. Optimizing approaches to deliver DNA efficiently into cells, and whether an adjuvant is necessary, are issues that are currently being explored.

It should be noted that as of mid-2020, there are no DNA or RNA vaccines that have previously been approved for human use.

Clinical Trials of Vaccine Candidates

The development of a new vaccine begins with what is called the discovery phase. During this time, a particular vaccine concept is chosen (e.g., inactivated virus, subunit vaccine, or RNA) and then tested in small animals, like mice. If the vaccine concept seems promising, it is tested in large animals. It is important that the chosen animal model recapitulates the symptoms and stages of the disease observed in humans. As monkeys are primates, they are similar to us in several ways, and so are often chosen to test vaccine concepts. Upon infection with the influenza virus, ferrets exhibit symptoms very similar to humans, thus they are often used to test concepts for influenza vaccines and therapies. During this stage of vaccine development, the data generated in animals are also used to estimate what may be a safe and effective dose for the vaccine. If all these preclinical studies go well, then the concept is ready for human clinical trials.

Clinical trials are carried out in three phases. In phase I, the safety of the vaccine is tested in a relatively small number of healthy humans. Based on animal studies, a range of doses is tested to determine the doses at which the vaccine can be administered without side effects. If the vaccine is proven to be safe in phase I, then in phase II of the clinical trial, a specific vaccine dose is chosen and the ability of the vaccine to elicit the desired immune responses is tested. For example, vaccinated people could be tested to determine if the vaccine elicited antibodies that can block the virus from infecting human cells. This is done by testing antibodies generated upon vaccination in the laboratory. It is also important to determine whether antibodies were generated in sufficient numbers to be protective. In the case of an ongoing pandemic, the level of antibodies required for protection from infection may not be known when clinical trials start. This was the case when clinical trials for COVID-19 vaccines started. Sometimes phases I and II of a clinical trial can be combined to accelerate vaccine development. As different vaccine doses are tested, the antibody levels and their neutralizing ability can be assessed at the same time.

After successful completion of phases I and II of the trial, the critically important phase III starts. This is when the efficacy of the vaccine in preventing infections in humans is tested. The gold standard of phase III clinical trials is what is called a double-blind trial. The people who enroll in the clinical trial are divided into two groups. One group is given the vaccine and the other is given a harmless substance (called a placebo). Neither the enrollees in the trial nor the physicians involved know who has received the vaccine and who is in the placebo group. Double-blind trials aim to minimize bias. Everyone enrolled in a phase III clinical trial is monitored for infection. The difference in infection rates and severity of illness between the group that received the vaccine and the placebo group determines the efficacy of the vaccine.

Phase III vaccine trials are enormously complex and expensive. When the efficacy of a new therapeutic for a disease like cancer is tested, everyone participating in the clinical trial has the disease. If there are 200 people enrolled in the trial, we assess the efficacy of the drug in a population of 200 people. When a vaccine is tested, at the beginning of the clinical trial, everyone enrolled is healthy. During the trial period, only a small fraction of these people will be exposed to the virus. Let's say that the natural prevalence of the disease that a virus causes is 2 percent. If 200 people are enrolled in the trial, only four individuals are likely to be infected during the course of the trial. So, the efficacy of the vaccine is really being tested in only four people. The statistical accuracy of the vaccine trial is therefore going to be much lower than that for the analogous cancer drug trial, even though both trials enrolled the same number of people. This is why phase III vaccine trials have to enroll very large numbers of people in both the group that gets the vaccine and the placebo group. By conducting clinical trials in areas with a high prevalence of the disease, the number of people enrolled in the trial can be reduced because more people are likely to be exposed to infection.

Many vaccine candidates that aim to prevent SARS-CoV-2 infection will be tested simultaneously in a short time. Enrolling sufficiently large numbers of people for each vaccine trial is a challenge. For this reason, some have wondered whether, given the urgent need for a vaccine that can prevent COVID-19, whether a different kind of trial called a "challenge trial" should be carried out. In such a trial, healthy volunteers are vaccinated and then infected with the virus. So, now, analogous to drug efficacy trials, everyone enrolled in the trial is infected. Thus, the efficacy of the vaccine can be determined with high statistical accuracy with small numbers of people enrolled in clinical trials. Since effective treatments for COVID-19 are not available, challenge trials present obvious ethical concerns.

Another critically important part of clinical trials and the deployment of a successful vaccine is manufacturing the product that goes into humans. Strict regulations exist for how such products are manufactured, which are called "good manufacturing practices" (GMP). In the United States, the Food and Drug Administration (FDA) regulates GMP standards in order to ensure the safety of pharmaceuticals. It usually takes a long time to develop the manufacturing process for a new drug or vaccine. This is especially true if there is no precedent for manufacturing a particular type of vaccine at large scale (as is the case with DNA or RNA vaccines). Typically, a manufacturing process involves many steps; each step has to be optimized, and methods to test the quality of the product have to be developed. The entire manufacturing process has to be developed before filing an application for approval by the FDA. Furthermore, as the clinical trial moves through different phases, the manufacturing process has to be scaled up. Changes to the core manufacturing processes at this stage requires a new application for approval, which can result in considerable delays.

With this general description of how vaccines work and how they are developed, let us first look to history and learn about how polio vaccines were developed. Then, we will describe some of the efforts underway to develop vaccines that may protect us from SARS-CoV-2 and HIV infections.

Examples of Vaccine Development

Salk, Sabin, and Polio Vaccines

The poliovirus has been circulating in humans for centuries, with evidence of polio found in Egyptian mummies. Polio is a very infectious food- and waterborne virus. Most infections result in mild flu-like symptoms or no symptoms at all. In a small proportion of cases (one in 200), infection causes muscle weakness or paralysis. In the twentieth century, outbreaks of polio began to occur with regularity. In many of the cases, children would lose the ability to walk or did not have the muscle strength to breathe. Interestingly, these outbreaks occurred more frequently in developed nations, like the United States and Scandinavian countries.

It is thought that before good sanitation and clean water supplies were available, most infants were likely to be infected with polio from contaminated water. Infants were able to control the virus because of protective antibodies acquired through their mother's milk. This mitigated the severity of disease in infants. The infants' immune responses also responded to the virus and so they acquired lifelong immunity. As hygiene and sanitation improved in some nations, infants were less likely to be infected by poliovirus. This was probably especially true for those born in wealthy families. After breastfeeding stopped, the mother's antibodies were no longer available for protection. So, people exposed to the virus at an older age had no immunity to it. It is thought that this increased the risk of severe disease and paralysis at older ages.

In 1916, a major epidemic occurred in the United States, with the epicenter being in New York City. Approximately 27,000 people fell ill, 6,000 died, and many children were paralyzed. As not much was known about the virus, and most infected people were asymptomatic, it seemed that randomly selected children suddenly became paralyzed. This situation was frightening. In 1921, Franklin Delano Roosevelt was infected with polio at age 39. Roosevelt had just lost the election for vice president of the United States. That a wealthy and powerful American politician could be afflicted by the disease and become paralyzed exacerbated the fear. Cases of paralysis from polio grew in number each summer, causing parents to dread summer vacation for their children. Many parents forbade their children from going to swimming pools, the beach, movie theaters, and bowling alleys. To address the national health crisis, Roosevelt started a philanthropic organization whose major goal was to develop a polio vaccine. This organization came to be called the March of Dimes.

Two New Yorkers and graduates of New York University medical school, Jonas Salk and Albert Sabin, would take two different approaches toward developing a vaccine for polio and in the process would become bitter rivals. The rivalry would pit those who believed in inactivated vaccines against those who believed in live attenuated vaccines. Salk and Sabin were not the first to advocate for these two different approaches for developing a polio vaccine. But when these strategies were first tested in the 1930s, clinical trials using both types of vaccines were believed to have caused polio. These events dampened enthusiasm for any further trials for the next 20 years.

As we described in the previous chapter, in 1949, Enders, Robbins, and Weller learned how to grow poliovirus in the laboratory. Salk immediately took advantage of this breakthrough. He scaled up production of the virus, and then determined just the right amount of formaldehyde required to inactivate it while keeping it intact. The March of Dimes decided to use all its resources to back the development of a polio vaccine based on Salk's inactivated virus. Given the national anxiety about polio, the US media focused on Salk's work. In a clinical trial, Salk was able to establish that his vaccine was safe, and also determine the vaccine dose required to elicit an antibody response. In three short years phases I and II of the trial were completed. In

1953, Salk announced that he was ready to test the efficacy of his vaccine.

The decision to start a large clinical trial was controversial. Enders and Sabin both questioned the safety of an inactivated vaccine, as well as whether an antibody response was a meaningful surrogate for protection from infection. Many clinicians also felt that a double-blind clinical trial was unethical as individuals in the placebo group would not benefit. Others were concerned that mostly children would be enrolled in the trial. Some worried that the wealthy would be more likely to volunteer their children since the disease afflicted them more, and this would bias the study. In the end, the phase III clinical trial used multiple approaches.

The clinical trial was an organizational tour de force. About two million children, almost all between 6 and 8 years of age, were enrolled. Since polio infections occurred mainly in the summer, all vaccinations needed to be completed before the end of the 1954 school year. The trial was conducted in counties with high rates of infection. One trial was a double-blind trial, with neither the physicians nor the children knowing which was the placebo group. In the other trial, the enrollees were first-, second-, and third-grade elementary school children. Only the second graders were vaccinated. Since no single company was able to manufacture the number of vaccine doses needed for the trial, many manufacturers were used. Vials of vaccines from different manufacturers were labeled similarly, but because there could be differences in product quality, the origin of each dose needed to be monitored.

At the end of the summer, the trial ended. Because of the massive amounts of data obtained, the computer company IBM was invited to help analyze the data. Finally, in the spring of 1955, on the tenth anniversary of Roosevelt's death, the March of

Dimes announced the exciting results of the trial, which showed that the Salk vaccine worked.

Almost ten years older than Jonas Salk, Albert Sabin greeted these results with mixed emotions. Sabin had been working on polio for almost his entire career. He was the one who recognized that polio infected the intestines first because of fecal contamination of food or water. After multiplying in the intestine, the virus then spreads to the blood before being cleared by the immune response. In some cases, the virus is able to enter the nervous system from the blood, resulting in paralysis. Based on this work, Sabin believed that a good vaccine needed to provide protective immunity to the intestinal tract.

Sabin spent years weakening or attenuating the poliovirus by growing it repeatedly in different animals and in cells in the laboratory. Eventually, he was able to isolate a weakened form of the virus that he felt was safe to use in humans. The United States and the March of Dimes felt that Salk's vaccine had solved the polio problem and there was no need for another vaccine. So, Sabin turned to other countries for support. In the Soviet Union, millions of people participated in a clinical trial. With its success established, the Soviet Union began manufacturing Sabin's vaccine. It is remarkable that at the height of the Cold War an American polio vaccine got its first foothold in the communist world. Eventually, Sabin's vaccine would be approved

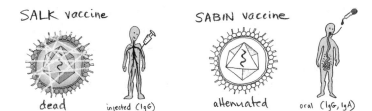

SALK vaccine

dead   injected (IgG)

SABIN vaccine

attenuated   oral (IgG, IgA)

for use in the United States in 1961 and, in a victory for Sabin, replaced Salk's vaccine in 1962.

Sabin's live attenuated virus vaccine had many advantages over Salk's vaccine. It was easier and cheaper to manufacture because, unlike Salk's vaccine, inactivation of the virus by careful treatment with formaldehyde was not required. Sabin's vaccine was also efficacious at lower doses, thus requiring fewer quantities to be manufactured. Administering Sabin's vaccine did not require syringes or needles. Vaccination involved swallowing a drop of fluid on a sugar cube. The live attenuated virus first infected the intestine, and elicited both IgG- and IgA-type antibodies. Recall from the immunity chapter that IgG protects the blood and IgA protects surfaces of organs. IgA antibodies induced by Sabin's vaccine prevented the poliovirus from attaching to and infecting intestinal cells. Salk's vaccine, in contrast, only induced an IgG response, so it was able to block the spread of the virus only in the blood and nervous system. If someone vaccinated with Salk's vaccine got infected, the virus could still infect the intestine and potentially spread to others via the feces. For reasons that are not completely understood, Sabin's vaccine induces lifetime immunity, while Salk's offered protection only for a few years. For these reasons, Sabin's vaccine soon became the standard vaccine used around the world.

Mainly because of the large-scale use of Sabin's vaccine, poliovirus has largely been eradicated from the planet. Only a few natural infections occur now, mainly in Pakistan and Afghanistan where polio is still endemic. Sabin's vaccine is a live RNA virus. While it does not thrive well in humans, it does replicate in us. Since RNA replication is error prone, the virus in the vaccine could mutate to become dangerous again. The mutated virus could spread to others and cause paralysis. Indeed, most of the cases of polio seen today outside the endemic areas are caused by such mutations of the live virus in Sabin's vaccine. Because of this, many countries, such as the United States, have returned to using Salk's vaccine as the standard method of childhood polio vaccination. In hindsight, it was fortunate that we had both the Salk and the Sabin vaccines. This is something to remember as we hope for a vaccine that can protect us from the COVID-19 disease.

The Race to a COVID-19 Vaccine

The human and economic carnage caused by the COVID-19 pandemic, the lack of effective therapies, and the realization that it would be difficult to acquire herd immunity through natural infection has led to many efforts to develop a vaccine. Here, we describe just two efforts as examples that we hope will illustrate the lightning speed with which COVID-19 vaccine development has proceeded.

Dan Barouch is a physician and a scientist at Harvard Medical School and the Ragon Institute in Boston. After finishing his education and postdoctoral training, he set up his own laboratory with the goal of developing molecular approaches to design vaccines that could halt global pandemics. His laboratory was primarily focused on developing an HIV vaccine using the adenovirus vectors that we described earlier. In 2007, Barouch and collaborators reported the creation of several such vectors. They inserted the genes corresponding to HIV proteins in one such vector, called Ad26, and clinical trials showed that the resulting vaccine was safe in humans and induced immune responses.

Barouch then launched a collaboration with the Johnson & Johnson (J&J) company to manufacture the HIV vaccine and perform a large clinical trial, efforts that are currently underway.

In 2016, when the Zika epidemic was going on, Barouch's lab inserted the genes of this virus into the Ad26 vector. Again with J&J, the vaccine was manufactured, and a clinical trial showed that just one dose of this vaccine resulted in neutralizing antibodies that were durable for a year. But, because Zika quickly waned in South America, a vaccine was no longer needed. Based on this work on Zika and HIV, Barouch and J&J established that the Ad26 vector was safe in humans and that millions of doses of a vaccine could be manufactured.

When the sequence of SARS-CoV-2 was made available on January 10, 2020, the Barouch laboratory was at its annual retreat. Because the sequence of the spike protein was distinct from other coronaviruses, the team realized that SARS-CoV-2 might cause a pandemic. On the evening of January 10, they started to design an Ad26-based vaccine that might protect against COVID-19. They focused on the spike protein, anticipating that neutralizing antibodies that could bind to its spike might be effective. In parallel, preparations were made for initiating studies in mice, monkeys, and ferrets. On January 25, Barouch called J&J, and four days later they had signed a collaboration agreement. J&J announced that they would invest $1 billion to make a billion doses of this vaccine if it proved successful in clinical trials. Animal tests subsequently showed that the vaccine elicited both antibody and T-cell responses. Clinical trials of this vaccine concept started in July 2020.

Moderna, a biotechnology company in Cambridge, Massachusetts, moved even faster. Moderna was established in 2010 and is focused on developing therapies and vaccines based on RNA delivery technologies. The SARS and MERS epidemics made clear that new coronaviruses could jump to humans, and were a serious threat. Scientists at Moderna had been studying how the new concept of RNA vaccines could be used to protect against such viruses. They found that a modified form of the RNA corresponding to the spike protein of MERS when encapsulated in a nanoparticle could elicit neutralizing antibodies to MERS in animals. Based on these studies, Moderna scientists were poised to help confront the COVID-19 pandemic.

Within 24 hours of the public release of the sequence of SARS-CoV-2, Moderna started work on creating an RNA vaccine that may protect against COVID-19. They quickly designed a modified form of the sequence of the RNA corresponding to the spike protein of SARS-CoV-2 and formulated a vaccine. While manufacturing an adenovirus-based vaccine takes some time, an RNA vaccine can be made in days to weeks. On March 16, in collaboration with the National Institutes of Health, Moderna started a phase I clinical trial. A press release shortly thereafter announced that the vaccine was safe in humans and could elicit antibodies against the virus's spike. Phase III efficacy trials could be completed in the autumn of 2020.

Neither an RNA or an adenovirus vaccine has ever been licensed for human use. So, if either type of vaccine succeeds, it will be a major achievement. Since the technology to produce adenovirus-based vaccines is more mature, mass production may be easier than for RNA vaccines.

Many other ideas and concepts for developing a COVID-19 vaccine are being pursued. For example, Oxford University and the company AstraZeneca are developing a vaccine based on concepts similar to, but distinct from, the J&J vaccine. This vaccine is also being tested in humans. Many other companies around the world are developing vaccines based on DNA and RNA, as well as other technologies. A collaboration between Pfizer and BioNTech is just one additional example. It is unclear at this time whether any of the vaccine concepts being pursued will be efficacious and whether the vaccine-induced immunity will be durable. So, it is wonderful that many vaccine ideas are being pursued in parallel. As the story of polio vaccines shows, having many vaccine options available can be very helpful. For COVID-19, it is almost essential that many vaccine ideas be pursued. If the virus continues to spread, we will need to vaccinate billions of people around the world to establish herd immunity. Having several efficacious vaccines available will help achieve this goal.

Clinical trials for vaccines that may protect us from COVID-19 are moving forward at a phenomenal pace. Manufacturing processes and facilities to make enormous numbers of doses of different vaccine candidates are being built in advance of knowing whether a particular vaccine will work. Some of these investments may well be for naught. But these facilities are being developed and built so that manufacturing does not become a bottleneck for rapidly deploying a successful vaccine. It is estimated that roughly 5 billion doses of COVID-19 vaccines will need to be deployed across the globe. If a booster shot is required, this number will double. In addition to manufacturing the vaccine, enormous numbers of vials to store and transport vaccines and devices like syringes necessary to inject vaccines into people will be required. These are enormous manufacturing and logistical challenges that we will have to overcome. Indeed, we will learn a lot from efforts to stop the COVID-19 pandemic.

The Long Pursuit of a HIV Vaccine

The HIV virus was identified over 35 years ago. Intense research efforts and significant financial investment since then has not led to a successful vaccine. This experience has made some wonder whether a vaccine that can protect us from COVID-19 can be developed in the near future. From a vaccine design perspective, however, HIV and SARS-CoV-2 are very different viruses. SARS-CoV-2 (and its spike protein) is not mutating very much. This suggests that if the current vaccines being developed can elicit neutralizing antibodies they will likely be efficacious.

In contrast, HIV is a highly mutable virus that replicates very quickly in humans. More than a billion new virus particles are generated in a single infected person every day, and many have mutations. Most of the mutant viruses are dysfunctional. But some are not. Our immune system mounts robust antibody and T-cell responses upon HIV infection. But very soon, mutant viruses emerge that are functional and can evade this response. For example, antibodies are produced that can bind to virus particles with a particular spike protein and prevent them from infecting new cells. After a few cycles of virus replication, mutants with different spike proteins emerge. The previously efficacious antibodies can no longer bind to the new spikes. Similarly, strong T-cell responses are mounted to HLA-bound fragments of HIV proteins. But soon strains with mutations in these protein fragments emerge. Of course, new immune responses are then mounted, but the virus mutates again. This is the reason why no one infected with HIV has been definitively identified to have completely cleared HIV infection. This also explains why there is no efficacious vaccine yet.

Progress is being made, however, to address the challenge presented by HIV's high mutability. Parts of the spike proteins of HIV do not mutate much. These regions need to stay the same for the virus to be able to infect human cells. If a vaccine could elicit antibodies that bind to these regions, it would prevent the vast majority of mutant strains of HIV from infecting our cells.

Antibodies that can achieve this goal are called broadly neutralizing antibodies (bnAbs). Some infected persons do produce
such antibodies, usually several years after infection, and not in
large enough numbers to clear the infection. But their existence
provides the proof of concept that the human immune system
can generate bnAbs. Intense efforts are underway to develop vaccines that can elicit bnAbs.

The other HIV proteins also have regions that cannot mutate
much. Interestingly, some combinations of mutations in HIV
proteins cannot emerge without making the virus dysfunctional.
If different killer T cells simultaneously attacked infected cells
that displayed the regions where combinations of mutations are
unviable, the virus would be trapped. The combination of mutations that could evade the T-cell responses would make the virus
dysfunctional, and if the mutations did not emerge, the killer
T cells would kill the infected cells. A small number of people,
called elite controllers, can mount T-cell responses that target
the mutational vulnerabilities of HIV. They do not eradicate the
virus from the body (with perhaps the exception of one very
recently reported case), but virus levels are very low in these people and so they do not proceed to the disease, or AIDS. Intense
efforts are underway to design vaccines that can elicit T-cell
responses that resemble those mounted by elite controllers.

Perhaps the work being done to develop vaccines that can
protect against diverse HIV mutant strains will one day be useful
for designing a vaccine that can protect us from all coronaviruses, past, present, and future.

Vaccine Safety

For most of us, the suffering caused by diseases like smallpox,
polio, mumps, rubella, and measles is just something we read

about in history books. This is because of the enormous success of vaccines in protecting the human population from many infectious disease-causing viruses. At the same time, fears that vaccines might cause disease have also been voiced. Recall from chapter 1 that only after variolation was shown to be safe in some prisoners and children did the Princess of Wales allow her children to be variolated. Similarly, in the early twentieth century, Henning Jacobson in Boston refused to be vaccinated against smallpox because he was concerned that the vaccine might make him sick. Such fears are not unreasonable because vaccination is a medical procedure performed on healthy persons, and involves injecting a foreign substance into an individual. But, over the years, vaccination has proven to be remarkably safe.

Early on, because of a manufacturing error, a batch of the Salk polio vaccine was not fully inactivated, and unfortunately resulted in polio infections. But modern manufacturing processes and strict regulation of GMP facilities make such errors unlikely today. In the United States, the FDA carefully reviews all data from preclinical and clinical trials before licensing a new vaccine for human use. The FDA and the CDC maintain the Vaccine Adverse Event Reporting System, and the CDC maintains the Vaccine Safety Datalink. These entities allow for careful, real-time tracking of the safety of vaccines. In the United States, there is also the National Vaccine Injury Compensation Program, which receives reports of adverse effects of vaccines. Between 2006 and 2013, for every million vaccine doses that were administered, there was roughly one report of an adverse reaction. Even the adverse effects due to mutations in Sabin's polio vaccine that have led to discontinuing its use in some places occur in only about one in a million vaccinated people.

An influenza vaccine in 1976 was linked to Guillain-Barre syndrome, which is a disease where the immune system attacks nerves. While this is scary, it now appears that such adverse events may be associated with viral infections and not the vaccine. Recent concerns about a connection between vaccination and autism are rooted in a 1998 paper in the journal *Lancet*, published by Andrew Wakefield and coworkers. In this work, the authors claimed that there was a connection between MMR vaccination and a "pervasive developmental disorder." Extensive follow-up studies have completely discredited this work, and the paper was retracted.

Adverse events can and do occur after vaccination, but these are rare events. Given this rarity, and the enormous benefit derived from protecting society from infectious diseases, it seems clear that vaccines have been a boon for humanity. Furthermore, noncompliance with vaccination programs puts vulnerable people in society at risk. The COVID-19 pandemic has made vivid the traumatic conditions that would prevail in a world without vaccines.

Epilogue

Since time immemorial, we have been at war with viruses. Innovations such as agriculture and the industrial revolution have improved the human condition. But these advances also led us to adopt lifestyles that favored the spread of highly contagious, infectious disease–causing viruses. We have largely been winning the war with these viruses because our immune systems can usually vanquish most viral infections, and because of technological innovations that have helped thwart the risk of contagion. These technological innovations include better sanitation, vaccines, and therapeutics. But we remain vulnerable for periods of time without protection when a new lethal virus emerges. Such events are inevitable because many viruses related to those that afflict us are also circulating in animals. Given the malleability of the genomes of RNA viruses in particular, new viruses can emerge that jump from animals to humans. When this happens, no one in the human population has protective immunity to it. If this virus is highly infectious and moderately lethal, devastating global pandemics can result.

The COVID-19 pandemic is a reminder that infectious disease–causing viruses are an existential threat to humanity. The costs of this pandemic have been enormous in terms of both human and economic damage. Almost a million people have died worldwide, tens of millions have lost their livelihoods, and global economic losses are measured in the tens of trillions of dollars. Governments are scrambling to respond, in many cases effectively limiting the death toll, but along the way creating a fiscal debt that will be borne by generations to come.

It doesn't have to be this way again. Informed by our history of battles with viruses, and the recent lessons learned from the COVID-19 pandemic, we need to create an integrated system of technologies that will help us prepare to respond more rapidly and effectively the next time, thereby saving millions of lives and trillions of dollars. A focused effort to create such a system will help us win the future by preparing for the "enemy" before it arrives.

We envision a future where a set of six interlinked technologies, developed based on rigorous and interdisciplinary scientific advances and informed by ethics and considerations of human well-being, can create a more pandemic-resilient world.

Early diagnosis and ongoing surveillance   It is absolutely essential that we be prepared to test for a new virus on a large scale. Based on existing advances in synthetic biology, nanotechnology, biosensing, and device engineering, we can create diagnostic tests that are reliable, portable, user-friendly, and minimally invasive. These diagnostic platforms will have interchangeable parts that can be rapidly adapted to test for a new virus or corresponding antibodies. With tools like this, we will be able to rapidly survey a large proportion of the population to accurately measure infection rate, monitor the spread of the infection in real time, classify potential patients to prompt appropriate follow-up actions, and help craft data-driven approaches to mitigate the spread of disease. Ongoing randomized testing of the population in a noninvasive and cost-effective manner will also allow real-time surveillance of new disease outbreaks and the fraction of people who have become immune.

The development of new virus surveillance and tracking tools will also pose many new ethical questions. Scientists, engineers, and public leaders will have to work very closely with humanists and social scientists so that the tools that help us fight pandemics do not do irreparable damage to free societies around the world. It is our hope that readers of this book who are humanists and social scientists will be inspired to be willing to work closely with scientists and engineers toward this end.

Developing robust epidemiological models Enormous amounts of diverse types of data are being collected during the COVID-19 pandemic. Modern computational power, advances in machine learning and artificial intelligence, and mechanistic understanding of immunology and epidemiology can be brought together to analyze these data and thus learn how viruses spread and the relative efficacy of mitigation strategies in ways that have not been possible heretofore. This knowledge will help create robust epidemiological models. Rigorous statistical analyses of the data obtained during the next pandemic using the types of testing devices described above will inform these improved epidemiological models, which can then make reliable projections to guide data-driven public health policy. In this regard, it will be especially important to combine epidemiological and economic data and models to assess the relative merits of mitigation strategies and suggest public health measures that can optimally balance public health needs with economic ones.

Vaccines SARS-CoV-2 is not mutating much as of now, but the challenges for vaccine design are much greater for pandemics caused by rapidly mutating viruses. By bringing together the life, physical, engineering, and medical sciences, we can create an approach for rational design of vaccines against highly mutable viruses. Sophisticated computational approaches can be applied to massive sets of data on virus sequences and structures and coupled to clinical data on disease progression. This approach can enable us to rapidly identify targets on the virus that a vaccine-induced immune response should hit in order to negate the virus's ability to mutate. These targets are regions of the virus's proteins where mutations are not tolerated because the mutant viruses are unviable. This knowledge can inform the design of vaccines that elicit immune responses against the mutationally vulnerable regions. Such a vaccine could protect against diverse strains of the mutating virus. This is because the targets that the vaccine-induced immune response would hit would corner the virus between a rock and a hard place—being killed by the immune responses or committing suicide by evolving mutations that evade the response. Such an approach could potentially result in universal vaccines against all coronaviruses, different influenza strains, HIV, and highly mutable viruses that may cause future pandemics. Successful development of such an influenza vaccine, for example, would require only one injection (or maybe a few, depending on the duration of protection) of the flu vaccine in your lifetime. This vaccine would confer protection against seasonally variant influenza strains and those that may cause a pandemic.

It is also possible that the knowledge gained from analyses of sequences, structures, and disease pathogenesis, combined with global virus surveillance capabilities enabled by the diagnostic devices noted earlier, may allow us to anticipate pandemic-causing viruses that may evolve in the future. This capability will allow the design of potential vaccines in advance of the emergence of a dangerous virus.

The vaccines being developed in response to the COVID-19 pandemic will provide us with important lessons regarding the efficacy of novel vaccine constructs, such as those that use RNA or DNA. Further progress in novel vaccine delivery modalities can lead to more efficient immunization strategies.

Social scientists, doctors, ethicists, scientists, engineers, and national leaders across the world will have to work together to clearly communicate the benefits of vaccination to society and describe why the associated risk is low. Without such programs, the development of sophisticated vaccine technologies will be for naught.

Antiviral therapeutics Therapies that can cure disease are game-changing. The rapid development of antiviral therapies tailored to the most vulnerable replication mechanisms of a new pandemic-causing virus can be achieved by leveraging strengths in artificial intelligence, bioengineering, and basic biology. Specifically, combining high-throughput screening of drugs and their viral targets with novel machine-learning approaches can enable rapid discovery, design, and development of new antiviral drugs. Artificial intelligence–based drug design augmented by lessons learned from past failed and successful drugs and new discoveries in virology can lead to designing bespoke anti-viral drugs tailored to a specific virus. Taken together, these technologies will enhance our arsenal of potential antiviral therapies.

Manufacturing Formulation and manufacturing of billions of doses of new biologic materials, either vaccines or therapeutics, usually takes many months because it is a formidable challenge. We are only beginning to understand how to develop flexible manufacturing methods that will enable us to rapidly make new products that are safe to administer to humans. We need advanced manufacturing approaches and compatible regulatory policies to enable large-scale manufacturing of vaccines and therapies to begin shortly after successful clinical trials. Absent this ability, during the COVID-19 pandemic, governments and philanthropists were investing in manufacturing infrastructure before vaccines were even in the late stages of clinical trials, thereby taking on significant financial risk. Lessons learned during the manufacturing of vaccines for COVID-19 will be helpful in shaping the future. The COVID-19 experience will also help us to optimally time and stage clinical trials for vaccines. We will also have to learn how to rapidly develop the infrastructure to make billions of doses of a vaccine, and mobilize the equipment and materials needed to store, transport, and deploy these vaccines.

Safer living spaces, workplaces, and hospitals Person-to-person transmission is an important factor determining the spread of infectious diseases. A fundamental understanding of transmission modalities and mechanisms, and their link to both host physiology and environments, is required to inform the design and retrofitting of housing, workplaces, and hospital environments to minimize transmission. This can be achieved by bringing together approaches from fluid dynamics, aerosol science, optical sensing, signal processing, virology, and monitoring of infected persons.

It is critical to recognize that our vision is not entirely "futuristic." The technologies described here are likely within reach with adequate investments to advance science and engineering. But getting there requires recognizing a fundamental barrier: many of the needed advances are not in the interest of individual companies who are uncertain whether investing in making these goods will be profitable absent another pandemic. A coordinated program of public-private-academic partnerships is required to undertake the necessary research, development, and manufacturing.

Governments must play a leadership role by signaling that they will support the purchase of pandemic preparation materials and by providing the support for basic research and development required to push forward the technological frontier. Academic institutions must provide leadership in answering the questions that broadly advance our scientific and technological knowledge and in working with private companies to turn those answers into reality. Private companies must respond to the signals coming from government and the knowledge coming out of academia with a coherent and productive plan for creating the goods that we need. Finally, private philanthropists and foundations can provide rapid and flexible funds to help establish the backbone of such an initiative.

It is is important to realize that students and young scientists join faculty members in universities to carry out the research that leads to new discoveries and inventions, such as the ones we envisage above. Therefore, investments in this research will necessarily result in the education of a generation of leaders who can create a more pandemic-resilient world. In the United States, after the launch of *Sputnik*, a generation of young people was inspired to pursue careers in science and engineering. Many societal benefits accrued from their work. Perhaps the same will happen after the COVID-19 pandemic.

History also teaches us that pursuing the goals we have articulated above will yield a high return to the world economy. In the United States, government support of research and development during and in the decades after World War II created the modern American economy—and enriched a generation of workers. There continues to be outsized returns to investments in basic science. As just one example, the US government's investment of $3 billion in the Human Genome Project has created 280,000 jobs, and the genomics sector pays more than $6 billion a year in US taxes. Investing in a pandemic-resilient future will ensure that the world creates more well-paid jobs while also advancing healthcare. This will require coordination between governments, private and academic sectors, and the healthcare delivery system. We anticipate that some of the scientific and technological advances that will emerge will become essential tools for healthcare unrelated to pandemics.

We hope that the development of new approaches to anticipate, prepare, and combat future pandemics will be pursued by all nations in cooperation. As the COVID-19 pandemic, and many that have come before, make clear, viruses do not discriminate between different peoples or respect national borders and walls. The human race is connected by our shared history of combating the same viruses, and we need to build the shields that will protect us all from these enemies in the future.

We know with certainty that there will be another pandemic. Whether it will be a further mutation of SARS-CoV-2 or some new pathogen, and when it will happen, is unclear. What is clear is that unless we make far-sighted investments in science, technology, and human capital, we will suffer needless deaths and economic catastrophe again. Let's make the investments required to realize the future that we describe here by the time we face our next pandemic. We hope that the readers of this book from across the spectrum of society will play an important role in the debates that are happening, and will happen, to see how best to win the future. Indeed, this was the sole purpose for writing this book.

Acknowledgments

We are indebted to Dr. Philip Stork for painstakingly drawing all the illustrations in this book. We have benefited from discussions with Abul Abbas, Paul Allen, Kristian Andersen, David Baltimore, Dan Barouch, Deepta Bhattacharya, Sara Cutler, Tony DeFranco, Michael Diamond, Yonatan Grad, Jonathan Gruber, Clifford Lowell, David Masopust, Bernhardt Trout, and Emil Unanue. Some aspects of the views expressed in the epilogue were based on discussions that led to an opinion piece written by AC and Jonathan Gruber. However, none of those noted above bear responsibility for the material presented in this book. We are also grateful to several others who commented on sections of this book. Detailed descriptions of the historical material presented in chapters 1 and 2 have been provided by many others in the past. We have especially learned from, and been inspired by, Arthur Silverstein's beautiful book, *A History of Immunology*. Finally, we express our sincere gratitude to our families, which put up with us while we worked at our "day jobs" and on this book during the lockdowns necessitated by the COVID-19 pandemic in the spring and summer of 2020.

Arup K. Chakraborty
Lexington, Massachusetts

Andrey S. Shaw
San Francisco, California

Suggested Reading

Chapters 1 and 2

Silverstein, A. M. *A History of Immunology*. San Diego: Academic Press, 1989.

Chapter 3

Andersen, K. G., A. Rambaut, W. I. Lipkin, E. C. Holmes, and R. F. Garry. "The Proximal Origin of SARS-CoV-2." *Nature Medicine* 26, no. 4 (2020): 450–452.

Lodish, H., Berk, A., Zipursky, S. L., Matsudaira, P., D. Baltimore, and J. Darnell. *Molecular Cell Biology*. 4th ed. New York: W. H. Freeman, 2000.

Lowen, A. C. "Constraints, Drivers, and Implications of Influenza A Virus Reassortment." *Annual Reviews of Virology* 4, no. 105 (2017): 105–121.

Chapter 4

Abbas, A. K., A. H. Lichtman, and S. Pillai. *Basic Immunology: Functions and Disorders of the Immune System*. 6th ed. Philadelphia: Elsevier, 2020.

Murphy, K., and C. Weaver. *Janeway's Immunobiology*. 9th ed. London: Garland Science, 2016.

Chapter 5

Ferretti, L., C. Wymant, M. Kendall, L. Zhao, A. Nurtay, L. Abeler-Dörner, M. Parker, D. Bonsall, and C. Fraser. "Quantifying SARS-CoV-2 Transmission Suggests Epidemic Control with Digital Contact Tracing." *Science* 368 (2020): eabb6936.

Hatchett, R. J., C. E. Mercher, and M. Lipsitch. "Public Health Interventions and Epidemic Intensity during the 1918 Influenza Pandemic." *Proceedings of the National Academy of Sciences USA* 104, no. 7582 (2007): 7582–7587.

Haushofer, J., and C. J. E. Metcalf. "Which Interventions Work Best in a Pandemic." *Science* 368, no. 1063 (2020): 1063–1065.

Kissler, S. M., C. Tedijanto, E. Goldstein, Y. Grad, and M. Lipsitch. "Projecting the Transmission Dynamics of SARS-CoV-2 through the Postpandemic Period." *Science* 368, no. 860 (2020): 860–868.

Metcalf, C. J. E., D. H. Morris, and S. Park. "Mathematical Models to Guide Pandemic Response." *Science* 369, no. 6502 (2020): 368–369.

Chapter 6

Flint, S. Jane, Vincent R. Racaniello, Glenn F. Rall, Anna Marie Skalka, and Lynn W. Enquist. *Principles of Virology*. 4th ed. Hoboken, NJ: Wiley, 2015.

Li, G., and E. De Clercq. "Therapeutic Options for the 2019 Novel Coronavirus (2019-nCoV)." *Nature Reviews Drug Discovery* 19, no. 149 (2020): 149–150.

Yin, W., C. Mao, X. Luan, D.-D. Shen, Q. Shen, H. Su, et al. "Structural Basis for Inhibition of the RNA-Dependent RNA Polymerase from SARS-CoV-2 by Remdesivir." *Science* 368, no. 1499 (2020): 1499–1504.

Chapter 7

Bloom, S., and I. Geesink. "A Brief History of Polio Vaccines." *Science* 288, no. 5471 (2000): 1593–1594.

Haynes, B. F., and D. R. Burton. "Developing an HIV Vaccine." *Science* 355, no. 6330 (2017): 1129–1130.

Hedrick, S. M. "The Imperative to Vaccinate." *Journal of Pediatrics* 201 (2018): 259–263.

Juskewitch, J. E., B. A. Carmen, B. A. Tapia, and A. J. Windebank. "Lessons from the Salk Polio Vaccine: Methods for and Risks of Rapid Translation." *Clinical and Translational Science* 3 (2010): 182–185.

Plotkin, S. "History of Vaccination." *Proceedings of the National Academy of Sciences (USA)* 111, no. 34 (2014): 12283–12287.

Sabin, A. B., and L. R. Boulger. "History of Sabin Attenuated Poliovirus Oral Live Vaccine Strains." *Journal of Biological Standardization* 1, no. 2 (1973): 115–118.

Index

Page numbers followed by an "f" indicate figures.